Geography Meets Gendlin

Janet Banfield

Geography Meets Gendlin

An Exploration of Disciplinary Potential through Artistic Practice

Janet Banfield
University of Oxford
Oxford, United Kingdom

ISBN 978-1-137-60439-2   ISBN 978-1-137-60440-8   (eBook)
DOI 10.1057/978-1-137-60440-8

Library of Congress Control Number: 2016954638

© The Editor(s) (if applicable) and The Author(s) 2016
This work is subject to copyright. All rights are solely and exclusively licensed by the Publisher, whether the whole or part of the material is concerned, specifically the rights of translation, reprinting, reuse of illustrations, recitation, broadcasting, reproduction on microfilms or in any other physical way, and transmission or information storage and retrieval, electronic adaptation, computer software, or by similar or dissimilar methodology now known or hereafter developed.
The use of general descriptive names, registered names, trademarks, service marks, etc. in this publication does not imply, even in the absence of a specific statement, that such names are exempt from the relevant protective laws and regulations and therefore free for general use. The publisher, the authors and the editors are safe to assume that the advice and information in this book are believed to be true and accurate at the date of publication. Neither the publisher nor the authors or the editors give a warranty, express or implied, with respect to the material contained herein or for any errors or omissions that may have been made.

Printed on acid-free paper

This Palgrave Macmillan imprint is published by Springer Nature
The registered company is Nature America Inc. New York

For Mum and Dad

Acknowledgements

First and foremost, I owe huge thanks to the talented, warm and generous individuals who gave of their time, experience and enthusiasm to participate in my research, and who welcomed me into their homes and artistic practices. It has been a pleasure and delight to work with Jane M, Laura, Katherine, Susan, Clare, Kassandra, Philippa, Yoko, Marnie, Polly, Ticia and Jane O, all of whom have been a source of inspiration. Thank you.

Thank you, also, to my supervisors, psychological and geographical. I thank Mark Burgess for his continued support, encouragement and guidance, despite my departure from psychology and return to geography for my doctorate, and in particular for his notification of a summer workshop through which I was introduced to the work of Eugene Gendlin. Without Mark's interest in my work, this book would not have come into being. I also owe thanks to Derek McCormack for his copious critical feedback during my doctorate. Thanks go, too, to my assessors and examiners, for their assistance in the development of my doctoral work. In particular, I am grateful to Jamie Lorimer and John Wylie for their continued support for my research interests, academic endeavours and career aspirations. Other much appreciated sources of support include Pam Berry for her support and encouragement during my time at Oxford, and to teachers at both primary and secondary levels. Specifically, thanks go to Mrs. Waylen, whose tales and artefacts from around the world first sparked my interest in geography, and to Mrs. Trim whose enthusiasm drew me to physical geography during my secondary school years. Their inspirational teaching, motivation and encouragement stimulated and nurtured an abiding fascination with geography.

I extend my thanks to an anonymous reviewer who has provided consistent support for the publication of my work on Gendlin in a variety of formats, and who has been formative in determining the focus of this volume. Thank you for sticking with it; I hope I have taken on board your valuable guidance effectively. I am also grateful to Rachel Krause Daniel and Elaine Fan at Palgrave Macmillan, whose enthusiasm and support have been hugely invigorating.

Finally, I am indebted to family and friends for their ceaseless encouragement and patience. Thank you, Mum and Dad, for the financial, practical and emotional support to help me get back on my feet (literally and metaphorically) and start new chapters in my life, both professionally and with "the feathers". Thank you, also, for your unwavering faith in me and for your tireless accommodation of my periodic niggles, anxieties and agitations. I would be lost without you. Darren, as ever, provides inspiration and support in a way only a big brother can, managing simultaneously to keep my head in the clouds and my feet on the ground, for which I am immensely grateful. Equally valuable in helping me to maintain both momentum and perspective has been the constancy of friendship; to Cerys, Jacqui, E-J, Kumiko, Clare, Shazia, Ruth, Hannah, Charlotte and Stu, Adrian and Karen, thank you. Special mentions go to Kumiko, Ruth, Clare and Cerys, for kindly and courageously offering to check the draft for intelligibility. Final responsibility, however, remains with me, so any outstanding errors, omissions or shortcomings are entirely my own.

CONTENTS

LIST OF FIGURES

Introduction

Most people, most of the time, do not give much thought to ideas about space and place. We might relax at home, or stroll in the countryside, or occasionally think about global issues, but we rarely have cause to stop and think about how ideas like home, countryside and the global have developed and how they influence our lives and experiences. Geographers, on the other hand, give much thought to these sorts of issues (McCormack 2008b): how we understand space; how we experience different places and make certain places meaningful; how different spaces and places influence our behaviour; and how we generate our own spaces through doing what we do. As a geographer, my particular interest is in spatial experiences of, and the spaces generated through, artistic practices. Experiences like that of one participant in my research (Laura), who described beginning to be in the scene she was painting even though she was physically confined to the studio in which she was working. Laura says "it's you, you're there, you're in it" and that the place in the painting "comes alive under your brush". Laura describes such experiences as immensely joyful, so these are not just incidental features of her artistic doing, but are personally meaningful, and the spatiality generated—the conjoined experience of her studio and the coastal scene—has tangible qualities. Far removed from most people's everyday ideas of space and place, how might geographers understand these kinds of spatial experience?

It is precisely these kinds of spatialities that I hoped to encounter and explore in my doctoral research into the emergence of spatiality (experience of space) and subjectivity (sense of self) in artistic practice, and from which this book has arisen. While it does not present a full account (for which, see Banfield 2014), the chapters that follow are strongly influenced by the research, and in places draw directly on my doctoral thesis. Chapter 7 contains the greatest amount of entirely new material, although all seven other chapters contain material that has been significantly reworked and developed, and incorporate a considerable amount of new material. Although based in a geography department, my research drew on academic interests in both geography and psychology. This book is a direct result of this interdisciplinarity, as it explores the potential for geography to benefit from the introduction of philosophical and methodological perspectives from psychology, particularly Eugene Gendlin's philosophical and psycho-therapeutic work, which is—as yet and as far as I am aware—unfamiliar to geography (Gendlin 1980, 1989, 1995, 1997, 2001, 2009a, b).

Gendlin's philosophical work falls within a stream of thinking known as *non-representational*. Rather than focusing on the representational content of texts, diagrams, maps, paintings and so on, and assuming that they passively and accurately refer to an external static reality, non-representational inquiry pays more attention to the processes that lead to the creation of representational forms and the influences that representations have in the world. In relation to a painting, for example, non-representational research would be more interested in the skills and moods utilized in the production of the painting and in the impact it has on those viewing it, than in the reality or truthfulness of the scene depicted. Non-representational thinking does not dismiss representation but adopts a different understanding of it: as productive rather than reproductive. Representation is treated as a practice not a product. As a result, non-representational inquiry is interested in diverse factors involved in the doing of representing, including embodied knowledge, intuition, emotions, sensibilities and dispositions. In Laura's experience of starting to be in her painting, the generation of a powerful spatial experience in the midst of the practice of creating the artwork is quintessentially non-representational.

While I could have drawn on numerous philosophies already familiar within non-representational geography, such as those of Henri Bergson and Gilles Deleuze (Greenhough 2010), or Jacques Derrida and Jean-Luc Nancy (Wylie 2010), I engaged instead with Eugene Gendlin, whose work has the potential to make a valuable contribution to the sub-discipline.

Gendlin acknowledges a number of philosophers whose thinking has informed his own, stating that without the work of Plato, Aristotle, Dewey, Husserl, Heidegger, Merleau-Ponty and others, he would not have been able to produce the work he did produce in quite the same way (Gendlin 2006).

Gendlin's philosophical work—*A Process Model* (2001)—addresses the relation between the reflective and the pre-reflective, or the representational and the non-representational, and seeks to think with more than conceptual structures, forms and distinctions (Gendlin 1989, 1993, 1995, 1997). In an autobiographical account (Gendlin 1989), Gendlin identifies himself with phenomenology, which emphasizes the interweaving of humans with their environment and promotes understanding the essence of things through our own embodied experience of them (Relph 1985; Merleau-Ponty 1995; Ingold 2011b). However, he says that he is able to work with differences between concepts in a way unavailable to phenomenologists, discarding mere descriptions and drawing attention to more-than-logical progressions between concepts. Gendlin asserts that statements can make mere logical sense or they can lift out more, giving more specificity and precision than logic alone (Gendlin 1989).

In addition to the *Process Model*, Gendlin has been at the forefront of the development of Experiential Psychotherapy, teaching at the University of Chicago for over thirty years, and founding and editing the journal *Psychotherapy: Theory, Research and Practice* (Gendlin 1989). Through his psychotherapeutic work he has developed a step-by-step training system for direct reference to the pre-reflective (which Gendlin calls the implicit), through which we can originate new meanings and define new concepts (Gendlin 2009b). Gendlin's book *Focusing* has been translated into seventeen languages, and he has been honoured four times by the American Psychological Association for his development of Experiential Psychotherapy (The Focusing Institute 2011).

Certain features of Gendlin's work suggest that it might hold promise in relation to particular challenges currently faced by non-representational geography. As Gendlin specifically describes his philosophy as a non-representational philosophy of the subject, it has the potential to contribute to disciplinary debates about the status of the human subject (Pile and Thrift 1995; Thrift 1996, 1997, 2008; Whatmore 1997, 2006; Nash 2000; Gendlin 2001; Dewsbury 2009; Pile 2010; Wylie 2010; Blackman 2010, 2012). Similarly, Gendlin's work might alleviate the methodological challenges posed by the suggested impossibility of working with the non-representational on its own terms (Massumi 1995, 2002; McCormack 2003, 2010; Bondi 2005; Anderson 2006; Blackman 2010; Blackman and Venn 2010; Clough 2010; Pile 2010), as it proposes psychotherapeutic techniques to generate conceptual understanding from pre-reflective understanding (Gendlin 1993, 1995, 2009b). My aim in drawing on Gendlin within my own research was to initiate the pursuit of these specific contributions to non-representational geography. It is this introduction of Gendlin to geography with which this book is primarily concerned, rather than the broader research through which this introduction was orchestrated.

Gendlin presents his philosophy not as a finished product but as a first attempt that works to some extent, and he grants "permission to use it in any form whatsoever, or argue with, do anything with it" (Gendlin 2006: 8). It is in the spirit of exploratory engagement that I take the first tentative steps towards a Gendlin-inflected geography. Consequently, this book is not intended to be either a comprehensive review of non-representational geography or a compendium of geographical research into art, and it is neither definitive nor exhaustive in its consideration of Gendlin's work or its potential connections and relevance to geography. Rather, it is intended as a targeted intervention into both non-representational geography and the geographies of artistic practice, with the specific points of connection, potentialities and implications identified being determined by the particularities of the interaction between Gendlin's work and my own research.

The book considers empirically and critically the potential for Gendlin to inform geographical thinking and methodology, and is laid out in three parts. These parts address, in turn: the interdisciplinary context for my research; Gendlin's conceptual content as I consider it to be relevant to non-representational geography and geographies of artistic practice; and my experimentation with Gendlin's specific techniques for accessing and articulating from pre-reflective experience in my research.

Part 1 introduces two significant domains of geographical inquiry, which I drew together in my research, to lay out the disciplinary terrain within which my research is situated. The first is non-representational geography, where the term non-representational can be considered in simplistic terms as a way of thinking about things which emphasizes practical and pre-reflective ways of knowing. The second is the geographies of artistic practice, which, in general terms, is concerned with the spaces and places created in, constituted by, and generated through artistic forms and practices.

In Chap. 1, I chart a brief account of the development and nature of non-representational geography, and highlight some of its key features and challenges. Specifically, I draw out non-representational geography's emphasis on pre-reflective and practical ways of knowing and the key concept of affect, and discuss contemporary concerns regarding the status of the human subject and how we might access affective or pre-reflective experience for academic purposes. Into this overview, I introduce Gendlin's work and highlight significant points of connection and divergence between Gendlin's work and non-representational geography. While my more thorough exposition of Gendlin's philosophy and its relevance to geography is undertaken in Part 2, and my practical exploration of his methods is presented in Part 3, in Part 1 I summarize his core ideas to give a sense of his work, "in a nutshell", in the context of non-representational geography.

In Chap. 2, my attention switches to the geographies of artistic practice. I describe in brief the evolution of geography's understanding of art from a descriptive and representational form to a performative and transformative practice, emphasizing the increasingly active understanding of art in geography and a growing interest in practice-based and collaborative research methods within the sub-discipline. This sub-disciplinary activity is situated within a broader context of methodological development across the social sciences, characterized as a coming together of ethnographic, arts-based and practice-based inquiry, which increasingly parallels non-representational geographical interests in practical and pre-reflective ways of knowing. Against this methodological background, I outline the aims and methods of my own research into the emergence of spatiality and subjectivity in artistic practice, through which I explored Gendlin's potential contribution to geography.

Part 2 engages in detail with key terms from Gendlin's philosophy and is structured around three themes: the role of the implicit (or pre-reflective) in human-environment relations; Gendlin's insistence that we are able to generate new conceptual knowledge from our implicit understanding (a process that Gendlin calls explication); and the idea of more-than-logical progressions between concepts that we can exploit to generate new conceptual knowledge.

In Chap. 3, I introduce core ideas from Gendlin's philosophical work and illustrate them through empirical data from my research to highlight their geographical relevance within the context of geographical interest in artistic practice. I focus on aspects of Gendlin's work that resonate with contemporary geographical interests, such as human–environment relations, time and space, and agency, and I explicitly address what I consider to be Gendlin's potential to inform non-representational geographical concerns regarding human subjectivity and agency.

Chapter 4 focuses on a key concern in non-representational geography; our capacity to access and apprehend implicit or affective (pre-reflective) aspects of our experience that are typically considered to be beyond representation. I introduce Gendlin's notion of explication: the process of generating formal concepts from our pre-reflective experience, which Gendlin calls the implicit. Gendlin considers that we can explicate from our implicit understanding, and in this chapter I consider the explication of implicit or pre-reflective understanding into words and images. I also examine Gendlin's idea of sharp concepts, which are both firmly rooted in implicit understanding and tightly tied into formal conceptual frameworks, in the context of both verbal and visual (linguistic and artistic) concepts. I discuss these ideas in relation to the use of narrative and symbolism in artistic practice to explore the potential for artistic practice to facilitate the explication of implicit understanding as proposed by Gendlin. I also address emerging concerns within geographies of artistic practice in relation to the need for geographers to develop greater capacity for thinking conceptually about images and image-making (Hawkins 2015), a thread which also runs through the following chapter.

Chapter 5 takes a detailed look at Gendlin's thinking with regard to progressions, as a process of progression of understanding from implicit to explicit, and as the more-than-logical connection between supposedly distinct formal concepts, which allows us to make sense of things even if they do not make logical sense. I relate the first of these to recent geographical efforts to rethink abstraction as a productive rather than reductive practice (McCormack 2008a, b, 2012), and, through the diagrammatic reformulation of the research narrative presented in the previous two chapters, I work through an example of Gendlin's progression between verbal and visual concepts. The second of these prompts the detailed exploration of another of Gendlin's concepts—crossing—through which the more-than-logical connection between two or more concepts becomes available for us to explicate new implicit meaning for formal conceptualization. I explore Gendlin's crossing between linguistic/verbal concepts in the context of geographical debates about scalar terminology, and between visual concepts in relation to two of my own paintings. Through these discussions I also revisit and reinforce my elucidation of some of Gendlin's key philosophical terms.

Part 3 attends in detail to Gendlin's explicatory methods and my own experimental attempts to apply them in modified form in geographical fieldwork, to explore our capacity to access and apprehend the implicit (affective) and to stimulate the methodological innovation necessary if non-representational geography is to deliver this capacity (McCormack 2003, 2010; Bondi 2005; Anderson 2006; Blackman and Venn 2010; Clough 2010; Lorimer 2010; Pile 2010; Blackman 2012).

Chapter 6 describes the ways in which I adopted and adapted Gendlin's therapeutic techniques within my research. It presents empirical material in relation to research experiences with two participants who were particularly informative, especially when considered as a pair, suggesting tantalizing potential for these methods to aid disciplinary efforts to understand pre-reflective aspects of experience. Subsequently, I address apparent contradictions within the practices and accounts generated through my research, relating these methodological outcomes back to Gendlin's philosophical ideas to develop a Gendlinian account of otherwise perplexing findings.

In Chap. 7, I undertake a critical examination of my modification and implementation of Gendlin's explicatory techniques as a challenge to my advocacy of the potential of Gendlin's work to geography up to this point. The chapter considers aspects of the research participants' demographic and practice characteristics and, more importantly, features of my research design that potentially undermine the enthusiasm with which we might welcome Gendlin to geography. Through this discussion, I also characterize particular aspects of my research design in Gendlinian terms, which both serves to bolster my research against some of the potential weaknesses identified, and brings methodological implications within and beyond geography. On the basis of this critical discussion, I propose that aspects of the research design originally perceived as problematic are more productively considered as perplexing conundrums that invite further interrogation in future geographical engagements with Gendlin.

Through these chapters I introduce Gendlin's ideas as I perceive them to be relevant to non-representational geography and to the geographies of artistic practice, and illustrate them with reference to my own empirical data and my own hobby artistic practice. I also identify particular issues or difficulties that arise in relation to aspects of Gendlin's work in the context of my own research, and in relation to my own research design in the context of my exploration of Gendlin's potential contribution to geography. I conclude by proposing that Gendlin's philosophical and psychotherapeutic writings have much to offer non-representational geography, including: providing new concepts and terminology for geographical engagement and adaptation; informing contemporary concerns with the capacity for non-representational geography to accommodate human subjectivity and agency; and answering questions concerning the ability to access and apprehend our implicit understanding. These potential benefits are not confined to non-representational geography. Within geographies of artistic practice, I suggest that Gendlin's work provides new ways for us to think about artistic practices and the spatialities and subjectivities that emerge through them, and alternative avenues into thinking conceptually about images and image-making. Within more-than-human geographies, I also suggest that research into human–nonhuman relations might be supported by Gendlin's philosophical ideas on objects and agency and by Gendlin-informed explicatory techniques. Within qualitative research more generally, I suggest that my application of Gendlin-informed interview techniques raises implications for video-based and arts-based research, and encourages the employment of multi-stage and multi-method research designs tailored to individual participants in research into affect. In particular, the variations in an individual's capacity for engaging with their pre-reflective experience when doing an activity, when viewing an activity, and when viewing themselves doing an activity, suggests a particular role for video-elicitation techniques in sustaining reflective awareness while simultaneously re-establishing pre-reflective experience, potentially optimizing our capacity to articulate affect.

I hope to have delivered a focused but detailed geographical engagement with Gendlin's work within the context of my research, as a modest but targeted intervention into non-representational geography and the geographies of artistic practice, and to entice further geographical engagement with Gendlin both within and beyond its (sub-)disciplinary homes.

References

Anderson, Ben 2006 Becoming and being hopeful: Towards a theory of affect. *Environment and Planning D: Society and Space* 24 733–752

Banfield, Janet 2014 *Towards a non-representational geography of artistic practice*. Unpublished doctoral thesis, University of Oxford, Forthcoming online: https://ora.ox.ac.uk:443/objects/uuid:dd12e1c4-f222-435b-adc0-c1bb68e4f4ac

Blackman, Lisa 2010 Embodying affect: Voice-hearing, telepathy, suggestion and modelling the non-conscious. *Body and Society* 16 163–192

Blackman, Lisa 2012 *Immaterial bodies: affect, embodiment, mediation*. SAGE, London

Blackman, Lisa & Venn, Couze 2010 Affect. *Body and Society* 16 7–28

Bondi, Liz 2005 Making connections and thinking through emotions: Between geography and psychotherapy. *Transactions of the Institute of British Geographers* 30 433–448

Clough, Patricia T 2010 Afterword: The future of affect studies. *Body and Society* 16 222–230

Dewsbury, John-David 2009 Affect. *in:* Thrift, N & Kitchin, R (eds.) *International encyclopedia of human geography.* Elsevier Science, Amsterdam; London 20–24

Gendlin, Eugene T 1980 Imagery is more powerful with focusing: theory and practice. *in:* Shorr, JE, Sobel, GE, Robin, P & Connella, JA (eds.) *Imagery. Its many dimensions and applications.* Plenum Press, New York; London 65–73

Gendlin, Eugene T 1989 Phenomenology as non-logical steps. *in:* Kaelin, EF & Schrag, CO (eds.) *American Phenomenology: origins and developments.* Kluwer, Dordrecht 404–410

Gendlin, Eugene T 1993 Words can say how they work. *in:* Crease, RP (ed.) Proceedings, Heidegger Conference. Stony Brook, State University of New York

Gendlin, Eugene T 1995 Crossing and dipping: Some terms for approaching the interface between natural understanding and logical formulation. *Minds and Machines* 5 547–560

Gendlin, Eugene T 1997 The responsive order: A new empiricism. *Man and World* 30 383–411

Gendlin, Eugene T 2001 *A Process Model* The Focusing Institute, New York

Gendlin, Eugene T 2006 Transcript of Gendlin Templeton Lecture. Psychology of Trust and Feeling Conference. Stony Brook University. http://www.focusing.org/gendlin_templeton.html. Accessed 05 Nov 2012

Gendlin, Eugene 2009a What first and third person processes really are. *Journal of Consciousness Studies* 16 332–362

Gendlin, Eugene T 2009b We can think with the implicit, as well as with fully-formed concepts. *in:* Leidlmair, K (ed.) *After cognitivism: a reassessment of cognitive science and philosophy.* Springer, London, New York 147–161

Greenhough, Beth 2010 Vitalist geographies: life and the more-than-human. *in:* Anderson, B & Harrison, P (eds.) *Taking-place: non-representational theories and geography.* Ashgate, Farnham 37–54

Hawkins, Harriet 2015 Creative geographic methods: knowing, representing, intervening. On composing place and page. *Cultural Geographies* 22 247–268

Ingold, Tim 2011b *The perception of the environment: essays on livelihood, dwelling and skill* Routledge, London

Lorimer, Jamie 2010 Moving image methodologies for more-than-human geographies. *Cultural Geographies* 17 237–258

Massumi, Brian 1995 The autonomy of affect. *Cultural Critique* 83–109

Massumi, Brian 2002 *Parables for the virtual: movement, affect, sensation* Duke University Press, Durham, NC; London

McCormack, Derek P 2003 An event of geographical ethics in spaces of affect. *Transactions of the Institute of British Geographers* 28 488–507

McCormack, Derek P 2008a Geographies for moving bodies: Thinking, dancing, spaces. *Geography Compass* 2 1822–1836

McCormack, Derek P 2008b Thinking-spaces for research-creation. *Inflexions* 1 1–15

McCormack, Derek 2010 Thinking in transition: the affirmative refrain of experience/experiment. *in:* Anderson, B & Harrison, P (eds.) *Taking-place: non-representational theories and geography*. Ashgate, Farnham 201–220

McCormack, Derek 2012 Geography and abstraction: Towards an affirmative critique. *Progress in Human Geography* 36 715–734

Merleau-Ponty, Maurice 1995 *Phenomenology of perception* Routledge, London

Nash, Catherine 2000 Performativity in practice: Some recent work in cultural geography. *Progress in Human Geography* 24 653–664

Pile, Steve 2010 Emotions and affect in recent human geography. *Transactions of the Institute of British Geographers* 35 5–20

Pile, Steve & Thrift, Nigel J 1995 *Mapping the subject: geographies of cultural transformation* Routledge, London

Relph, Edward 1985 Geographical experiences and being-in-the-world: the phenomenological origins of geography. *in:* Seamon, D & Mugerauer, R (eds.) *Dwelling, place and environment: towards a phenomenology of person and world*. Columbia University Press, New York, Oxford 15–32

The Focusing Institute 2011 www.focusing.org. Accessed 07 Aug 2011

Thrift, Nigel J 1996 *Spatial formations* SAGE, London

Thrift, Nigel J 1997 The still point. *in:* Pile, S & Keith, M (eds.) *Geographies of resistance*. Routledge, London 124–151

Thrift, Nigel J 2008 *Non-representational theory: space, politics, affect* Routledge, London; New York

Whatmore, Sarah 1997 Dissecting the autonomous self: Hybrid cartographies for a relational ethics. *Environment and Planning D: Society & Space* 15 37–53

Whatmore, Sarah 2006 Materialist returns: practising cultural geography in and for a more-than-human world. *Cultural Geographies* 13 600–609

Wylie, John 2010 Non-representational subjects? *in:* Anderson, B & Harrison, P (eds.) *Taking-place: non-representational theories and geography*. Ashgate, Farnham 99–114

Disciplinary Terrain and Connections

CHAPTER 1

Non-Representational Interest in Affect

Abstract In Chap. 1, Janet Banfield helpfully situates Eugene Gendlin's philosophical and psychotherapeutic work within the context of non-representational geography, highlighting consistencies and discrepancies concerning the relation between the reflective (representational) and the pre-reflective (non-representational), and the specific relation between affect, emotion and cognition. Banfield highlights some of the key challenges currently faced by non-representational geography—concerning the status of the human subject in an anti-essentialist philosophy assumed to deny a psychological subject, and our capacity for accessing and apprehending affect if the affective is assumed to escape representational capture—and specifies how a Gendlinian perspective might inform these concerns as a result of the different way in which Gendlin considers affect, emotion and cognition to be related.

INTRODUCTION

This chapter contextualizes Gendlin's philosophical and psychotherapeutic work within non-representational geography. It provides a brief account of the nature of non-representational geography, its key features and interests, and some of the critical challenges it faces. In particular, this introductory discussion of Gendlin's work is oriented around how Gendlin understands the relation between the representational and the non-representational,

© The Author(s) 2016
J. Banfield, *Geography Meets Gendlin*,
DOI 10.1057/978-1-137-60440-8_1

and the relation between his own term for the pre-reflective (the implicit) and that more familiar to non-representational geography (affect). Rather than attempting an exhaustive review of non-representational geography and the turn to affect (for fuller accounts of which see, for example, Thrift 2004a; Anderson 2006; Dewsbury 2009, 2010b; Anderson and Harrison 2010b; Wetherell 2012), this chapter provides a general sense of how Gendlin's work relates to specific non-representational geographical themes, activities and concerns, and the potential within it to invigorate contemporary debates and challenges, the exploration of which forms the substance of Parts 2 and 3.

Non-Representational Geography

The increase in humanistic approaches stimulated by disciplinary disaffection with positivist methods (Samuels 1978; Daniels 1985; Livingstone 1992; Rose 1993) put human experience back as a central concern to geography (Crang 1998). This encouraged detailed engagement with phenomenology, which seeks to reclaim direct primitive contact with the world to understand things in their essence, through our own embodied experience rather than presupposing scientific knowledge (Tuan 1971; Merleau-Ponty 1995; Parry and Wrathall 2011). Rather than thinking of self-contained humans distinct from the world, phenomenology proposes that the basic state of human existence is one of being-in-the-world, or actively dwelling within and interweaving with the environment (Tuan 1971; Merleau-Ponty 1995; Wylie 2006; Ingold 2011b). Buttimer and Tuan provide examples of the development of phenomenological work in geography, although more recently non-representational geography has started reworking certain phenomenological ideas, in ways discussed later in this chapter (Tuan 1971; Buttimer 1976; Wylie 2006).

Non-representational geography emerged in the mid 1990s (Thrift 2008; Anderson and Harrison 2010a) as concern grew to resolve the tension between the material and the symbolic in a reaction to a perceived stress on language and neglect of the materiality of the human body (Daniels 2004; McCormack 2004; Oakes and Price 2008; Thrift 2008; Anderson and Harrison 2010a; Leys 2011). Nigel Thrift's *Spatial Formations* (1996) has been identified as marking the inception of non-representational geography (Lorimer 2007), which favours the practical, processual and eventful over representation and collective symbolism, on the basis that we come to know things through active experience rather than passive observation (Anderson and Harrison 2010a; Greenhough 2010). Although a diverse and difficult body of work to summarize, non-representational thinking can be characterized in relation to two disciplinary turns: a performative turn; and a turn to affect. These turns are encapsulated by two features of non-representational geography: its insistence on not prioritizing representations as epistemological vehicles; and its valorization of processes that operate before conscious reflection (McCormack 2005).

As part of the performative turn in geography, non-representational thinking encourages a change in focus from systems and forms of representation to processes of practice and performance, as the production of knowledge is now framed not in the representation of an external reality but in the practical process of doing things (Dewsbury and Naylor 2002; Dirksmeier and Helbrecht 2008). Non-representational geography does not dispense with representations but reanimates them as active and affective interventions (McCormack 2005). With representations seen not in opposition to practices but generated through them (Driver 2003), the focus of non-representational geography is on how rather than what (Dewsbury 2010a). Here, thought is conceived as an intervention in the world, such that if we act differently we can think differently. As a result, thinking is more than cognitive (Thrift 2004b).

Described as a turn to affect, the growth of non-representational geography has brought an expansion and intensification of interest in pre-reflective registers of experience, such as our moods, dispositions, emotions, habits and capabilities (Dewsbury 2009; Blackman and Venn 2010; Blackman 2010, 2012; Featherstone 2010; Pile 2010; Leys 2011). Non-representational theories share a concern for the sensate and (post)-phenomenological dimensions of existence (Bissell 2010), emphasizing affective sensibilities as active (McCormack 2003; Wylie 2010). In contrast to the conventional phenomenological approach of positioning a body in a landscape and positing a self inside a body (Wylie 2005), non-representational theorists have reinvented much of phenomenology on the grounds that the world is not static (Thrift et al. 2010). Rather than determining essences, which formed the focus of phenomenology, non-representational thinking is concerned with uncovering conditions and processes of their emergence (Tuan 1971; Pels et al. 2002; Simpson 2009). The subject, too, is not essential but emergent; not a cognitive construct but a practical outcome. Subjectivity is neither static nor contained within the body, but distributed beyond it, and is constituted as much by moods and sensations as it is by ideas and beliefs. In many ways, Gendlin's work is consistent with these disciplinary developments, as it focuses on our pre-reflective experience, it is concerned with how we can articulate from that pre-reflective experience, and it provides a distributed, emergent and practice-based notion of subjectivity. Consequently, the rise of non-representational thinking in geography has brought about parallel shifts in focus from discourse to practice and from meaning to affect (Whatmore 2006), providing opportunities for geographical engagement with, and development of, Gendlin's work.

AFFECT, EMOTION, COGNITION

Within geography, affect is a contested term and is used in divergent ways (Anderson 2006), for example: as a sense of push, stimulus or compulsion in the world (Thrift 2004a); as a universal interconnectedness, intensity or process (Massumi 1995; Game 2001; Blackman 2012); or a pre-personal force or a capacity to affect and be affected (Anderson 2006; Bertelsen and Murphie 2009; Dewsbury 2009). In particular, the discipline accommodates divergent understandings of the relation between affect and emotion, and between affect/emotion and cognition. These issues provide my point of entry to Gendlin's work in relation to non-representational geography. I draw primarily on Brian Massumi's work on movement and affect (Massumi 1995, 2002) to provide a basis of comparison between Gendlin's work and philosophies more familiar to non-representational geography. Although other philosophers—for example, Spinoza, Bergson, Deleuze and Guattari—are also pressed into service within non-representational geography, I have prioritized Massumi for two reasons. One is that Massumi's work often features alongside these other philosophers, and has itself been influenced by them; the other is that Massumi's specific terminology resonates particularly well with Gendlin's, as I outline at key points. While I assume a certain level of understanding with regard to non-representational thinking in this book, detailed knowledge of these other texts is not necessary to understand the substance of Gendlin's work. These connections are intended to provide conceptual moorings in texts which might be more familiar, and to help situate Gendlin's work within a broader non-representational literature. These connections, then, can either be explored or ignored, depending on interest and preference.

Both affect and emotion are subject to multiple definitions (Anderson 2013), with the relation between the two terms being both key and contentious (Dewsbury 2009). For some, emotion is a tangible manifestation of affect, an expression or qualification of the affective (Thrift 2004a; Bertelsen and Murphie 2009; Dewsbury 2009); while for others affect is broader than emotion, neither reducible to it nor interchangeable with it, and emotion is related not only to affect but also to cognition as an ideological attempt to understand the affective (McCormack 2003; Grossberg 2009; Blackman 2012). For yet others, affect and emotion are different. Based in large part upon an engagement with Massumi's writings, some consider that affect and emotion operate according to different logics (Massumi 1995, 2002; McCormack 2006; Leys 2011). Understood in this way, a distinction is drawn between affect—which allows us to consider the human in more-than-human terms because affect is something like an atmosphere that is not confined to the human body—and emotion—which does not allow for such more-than-human thinking because emotion is personalized within the body. In other words, emotion lacks affect's transversality because it is specific to an individual human (Anderson and Harrison 2006; McCormack 2006).

Gendlin's term for affect is the *implicit*. The implicit fits the description as a pre-personal force or transversal interconnectedness, but it encompasses a different relation between affect and emotion. The implicit is pre-reflective and includes—but is not co-extensive with—emotion. Gendlin uses the term *felt sense*, and specifies that feelings are not just emotions but whole complexities that are felt in the body as an implicit consciousness. This is Gendlin's *implicit understanding*: a holistic feeling of the entire context (Gendlin 1980, 1989, 2009a). As examples, Gendlin says that the implicit lets us grasp the particular way in which a phrase with multiple meanings is intended, and that it lets us know we have forgotten something. It also lets us know when we have remembered something and, by extension, when the thing we have just remembered is not the thing that we originally forgot (Gendlin 1993). We can think of the implicit as the feeling of emotion, but also as connected with sensory experience, visceral sensations of bodily functions and proprioception, and "spooky" sensations of gut feeling, intuition and sixth sense, which all come together in the felt sense of a situation. The implicit, then, is in one sense broader than affect in that it incorporates rather than isolates emotion, but in another sense is potentially more internally differentiated than affect, although this conceptual differentiation does not preclude the cross-activation of emotional, visceral and spooky felt senses in lived experience. Consequently, Gendlin offers potential to inform efforts to distinguish affect from other non-semantic aspects of experience to reduce the breadth of current definitions (Grossberg 2009; Blackman 2012).

Sub-disciplinary domains which preferentially emphasize affect or emotion also differ in their understandings of the relation between affect/emotion and cognition. Affect is assumed to be prior to meaning and below conscious awareness, non-symbolic and incapable of analysis in a semiotic mode; affect is said to be the matter and practice of our being, rather than that with which we understand that being (Dewsbury 2009; Gibbs 2009; Pile 2010; Leys 2011; Dixon and Straughan 2013). On this affectual reading, emotions are qualified into semantic and semiotic meanings through a process of capture and closure (Anderson 2006; Dewsbury 2009). The pre-reflective cannot be translated into reflective, conscious or cognitive understanding without inflicting violence on the totality of the original experience (Gibbs 2009). Cognitive understandings are reduced forms of pre-reflective understandings, and the pre-reflective or affective always exceeds its conceptual capture. Importantly, distinctions have been drawn between affectual geographies, which consider it impossible to trace back from thoughts to unqualified affects, and emotional geographies, which are deemed to allow for the reaching back of thought to the pre-reflective. Emotional geographies stress the significance of expressed emotions, whereas affectual geographies emphasize the importance of the inexpressible (Pile 2010). Although this distinction has been criticized for imposing an unwarranted separation of mind and body in affectual geographies, the resistance within non-representational geography to the possibility of tracing back from thought to affect does perhaps suggest some form of disconnection preventing us from articulating affect, even if it is not associated with ideas of mind and body (McCormack 2003; Pile 2010; Curti et al. 2011).

For Massumi, for example, the autonomy of affect is its openness, its perpetual escape; affect will remain beyond analysis unless a vocabulary can be found for that which is imperceptible and which cannot be found because it perpetually escapes perception (Massumi 2002). In identifying a gap between content and affect (Massumi 1995), Massumi seemingly precludes any connectivity between the two, which might otherwise allow for the conceptualization of affect. Although the postulation of a resonant relationship between content and affect goes some way to establishing a connectivity, the personal fixing of a quality of experience as emotion or cognition indicates the capture and closure of affect, which can never be complete because of its perpetual escape (Massumi 1995, 2002).

Complicating this situation, though, and providing a link with Gendlin's work, Massumi also talks about affect being the interface between implicate and explicate orders, using the same term for that which perpetually escapes and for the connective means by which it is (partially) captured (Massumi 1995). If, as Massumi proposes both thought and affect can extend to any level providing we account for their unique functioning at each level, if they have a resonant connectivity, and if affect itself provides an interface between the implicit and the explicit (Massumi 1995, 2002), then the question re-arises as to why we are deemed incapable of generating explicit concepts from implicit experience without losing their affective quality. The proposition that affect's creative potential is arrested or nullified when it is quantified or qualified on a personal level (McCormack 2003; Pile 2010) seems to me to sit uneasily in a field of inquiry that valorizes the affective and performative power of representation. There seems to be an inconsistency between the non-representational commitment to the affective power of representations in general and the denial of affective power within specifically personal or emotional representations.

By contrast, for Gendlin, all concepts bring with them more than their conceptual patterns because logical patterns do not work only logically, and we can enter that "more-than" which functions implicitly (Gendlin 1993). Gendlin proposes not only that we can think with the implicit, but also that we can do so deliberately (Gendlin 1993, 1997). Gendlin asserts that arguments that the more-than-logical is simply ineffable assume that language is conceptually structured, and that this assumption has been the biggest challenge for twentieth-century philosophy (Gendlin 1993, 1997). Gendlin overcomes this problem by proposing that concepts involve what they are about, that concepts are more-than-conceptual, as they both say and exemplify how their logical structure is exceeded (Gendlin 1993, 1997). For Gendlin, the issue is not that the affective or implicit *cannot* be rendered logically but that they cannot *yet* be rendered logically (Gendlin 1995). Like Massumi, Gendlin considers that what was implicit is changed in its explication. However, this is not a loss or alteration of the implicit but its "carrying forwards" by the words into which it is explicated (Gendlin 1993, 1995, 1997, 2009b). This carrying forward is Gendlin's non-representational connection between implicit and explicit orders, enabling us to keep a concept connected with its implicit origin, to re-enter and return with something further (Gendlin 2009b).

Gendlin's philosophy, then, allows for greater connectivity between reflective and pre-reflective modes of understanding, and for thinking with the implicit (Gendlin 1980, 1993, 2009a, b). Gendlin has developed specific techniques for articulating, which he terms explicating, from the implicit to develop new affectively attuned conceptual knowledge and to use our existing conceptual understanding to gain access to the affective (Gendlin 1993, 1995, 2009b). In this sense, Gendlin's philosophy is more closely aligned to Pile's characterization of emotional than affectual geographies (Pile 2010) in allowing us to trace back from the reflective to the pre-reflective, although its vocabulary is consistent with affectual rather than emotional geographies. Gendlin, then, offers potential to answer Pile's call for affectual models that allow for the emergence of affects into conscious or cognitive apprehension (Pile 2010).

For Gendlin, the creative potential of affect need not be arrested by its explication, but can be recharged and potentialized for further affective activation, in a manner seemingly more in keeping with the values and priorities of non-representational geography. While the expression is not the same as the experience, there remains connectivity between the two, to the extent that discourse is enlivened rather than rubbished (Wetherell 2012). For Gendlin, then, there is no paradox or hypocrisy in thinking the non-representational on its own terms (Bondi 2005; Pile 2010). We are not condemned to surrender to stultifying concepts, because we have the capacity to generate new concepts that are implicitly charged, meaning that we can talk about that which is more-than-conceptual in conceptual terms because those concepts are themselves more-than-conceptual (Gendlin 1993).

The dissolution of this paradox is possible because of Gendlin's understanding of the relation between the implicit and the explicit. The explicit (for example language) originally emerged from the implicit and remains connected to that implicit from which it emerged. In this sense, the implicit is prior to the explicit and can function independently of it. In contrast to a distinction identified earlier between the matter and practice of our being and the means by which we understand it (Dewsbury 2009; Gibbs 2009; Pile 2010; Leys 2011; Dixon and Straughan 2013), the implicit is both the stuff of our being and a non-cognitive way of understanding that being. The implicit might emerge into the explicit, but it does not necessarily do so, operating pre-reflectively rather than reflectively. By contrast, the explicit cannot function independently of the implicit, because even if we are not aware of the connectivity between the explicit and the implicit, that connectivity persists nonetheless. For Gendlin, we have become insensitive to the implicit by virtue of our reliance on linguistic convention and conceptual rather than implicit meaning. We use linguistic terms *as if* they were independent of their original implicit meaning, but that perceived independence is false.

This presents an interesting twist to the false consciousness that Pile identifies in McCormack's work with Dance Movement Therapy, in which the therapeutic practice was not founded on talking about feelings. Asking people about their feelings is proposed to prompt the articulation of socially scripted responses rather than immediately felt, embodied sensations, due to the inability to access affective experience (McCormack 2003; Pile 2010). For Gendlin, it is this supposed inaccessibility of the implicit and the assumed independence of conceptual understanding from implicit understanding that constitutes the false consciousness. We falsely assume that we cannot access our implicit understanding because we have become unaccustomed to doing so. It is on the basis of this yoking of the explicit to the implicit, but not of the implicit to the explicit, that I prefer the term non-representational, to the term more-than-representational, which is favoured by some authors (for example, see Lorimer 2005). The term more-than-representational seems to suggest that representation comes first and that the more-than is somehow supplementary to it, whereas the term non-representational seems more appropriate because it emphasizes Gendlin's insistence that the implicit can work independently of the explicit, but not vice versa.

For Gendlin, this capacity to generate implicitly charged conceptual terms can be cultivated, and he has developed specific psychotherapeutic techniques to help us generate such terms. Gendlin's psychotherapeutic work is not that dissimilar to ideas of becoming responsive to different surfaces of attention (McCormack 2003), but this becoming responsive is just one side of Gendlin's approach, with the other being the explication of that responsiveness into conceptual forms that remain implicitly charged, in a manner commonly denied in contemporary non-representational geography.

METHODOLOGICAL CHALLENGES

A significant and well-rehearsed challenge facing non-representational geography is its lack of a conceptual language with respect to the affective (Massumi 1995; Anderson 2006; Blackman 2010; Blackman and Venn 2010). Although such a vocabulary is gradually emerging, further work is still required to develop terminology to describe that which is commonly deemed irreducible to discursive orders, to understand how verbal expressions shape emotional experience, and to overcome problematic terms for intersubjective affectivity through transmission, circulation and contagion (Anderson 2006, 2013; Anderson and Harrison 2006; Pile 2010). Gendlin's work introduces a substantial and distinctive vocabulary, which we can use as a toolkit to supplement or modify existing geographical terms. In addition, his particular psychotherapeutic methods, in facilitating the generation of affectively attuned vocabularies, might also allow the development of new linguistic forms that avoid problematic associations and masculinist, technocratic metaphors considered to burden geographical discourse (Thien 2005; Grossberg 2009; Pile 2010). While in many instances it is not the words that count but the struggle to articulate that makes the incommunicable apparent (Dewsbury 2009), Gendlin suggests through his explicatory techniques the potential to generate words that do count, a suggestion that is considered further in Chaps. 4 and 6. Gendlin's techniques also have the potential to supplement existing disciplinary habits of hyphenating terms in the effort to accommodate dualistic concepts. A Gendlinian interweaving of two terms functions at both pre-reflective and conceptual levels, tracing back from their conceptual formality to their affective origins, to which we have largely become insensitive. This is addressed in Chap. 5.

Another and related challenge for non-representational geography is the need for methodological innovation and invigoration to access felt worlds as, despite substantial development of theoretical thinking, the development of non-representationally informed methods is less well advanced (Blackman and Venn 2010; Clough 2010; Lorimer 2010; Pile 2010). As affect is difficult to capture with conventional methodology, new approaches have been encouraged which explore the practices through which meaning emerges, and in which the researcher acts not as a neutral observer but as an active influencer, directing participants to that which remains unsaid (Blackman and Venn 2010; Blackman 2012; Anderson 2013). In particular, the synaesthetic and kinaesthetic nature of visual practices and social technologies, such as video, are attracting increasing attention in efforts to understand how such practices and technologies function affectively as intimate forms of knowledge (Thrift 2004a; Gibbs 2009; Blackman and Venn 2010; Featherstone 2010; Blackman 2012). Gendlin's psychotherapeutic techniques also offer potential here, as a possible means to apprehend or explicate affect, which can be applied in verbal, visual and practice-based research settings. Gendlin's techniques are predominantly verbal and are designed to help with the linguistic expression of deeply felt emotions and sensations we would normally consider beyond words. However, if visual practices and social technologies also afford access to experiences that are difficult to articulate, there might be considerable potential in combining visual practices, social technologies and Gendlin-informed techniques.

There are two core techniques within Gendlin's explicatory process: one is directed towards accessing our implicit or pre-reflective understanding (focusing); while the other is more concerned with integrating this understanding into formal conceptual frameworks of knowledge (thinking-at-the-edge). These are supplemented by a further process (dipping), which involves repeatedly returning to our implicit understanding to refine the implicit meaning of the conceptual terms we are generating. Gendlin proposes that we can access the implicit through focusing, which is a kind of body-centred intentional attunement to experience below cognitive or reflective awareness (Gendlin 1980). Through focusing, we allow a felt sense to form, making the implicit available for explication into formal conceptual knowledge. This emergent implicit understanding is integrated into our existing conceptual frameworks through thinking-at-the-edge, which aims to forge formal or logical connections between the emergent implicit understanding and our pre-existing explicit understanding, enabling us to say that which we could not previously say (Gendlin 2009a). Following on from this, Gendlin proposes that we can also use our existing concepts to trace back into our implicit understanding and re-invigorate our seemingly independent conceptual terms with implicit meaning (Gendlin 2009b).

Focusing is about making our implicit understanding available for explication; thinking-at-the-edge is about working across the edge between our explicit and our implicit understanding, to refine our emergent concepts. This involves the third practice—dipping—through which we return periodically to our felt sense to check our communication or conceptualization against it (Gendlin 1995). In this way, Gendlin considers that we can generate fresh ways of speaking, which carry forwards the felt sense, recognizable as a distinctly sensed change in the body (Gendlin 2009a). For example, we might write down an account of a certain felt sense and then identify key terms within it and consider how appropriate they really are by asking what it is that we really want these terms to mean for our felt sense, repeatedly returning to that felt sense to establish a resonance between the implicit and explicit meanings of the terms we generate.

Importantly, Gendlin stresses the need to let each thing be and be heard just as it emerges, and not to force emergent terms into conceptual or linguistic frameworks prematurely. By freeing ourselves of linguistic convention as new utterances emerge, we have the potential to generate totally new utterances, terms and concepts, which can then be progressively and more meaningfully tied into conceptual frameworks (Gendlin 2009a).

Chapter 6 details my own experimental attempt to integrate three different notions of intimate knowledge: Gendlin's *explicatory methods*, which are applied through *artistic practice* as its own form of focusing, and in association with *video-elicitation techniques* in an attempt to re-establish retrospectively the bodily felt sense of that artistic practice. This experimental fieldwork is subsequently critiqued in Chap. 7, in preparation for bringing the work to a conclusion.

Subjective Challenges

As well as being post-phenomenological, non-representational thinking is post-human and post-structural as the subject is seen as neither an inherent and unchanging personal attribute, nor the passive outcome of social structures and influences. With the non-representational retreat from an intentional psychological subject in favour of emergent subjectivities (Dewsbury 2009; Pile 2010), the geographic self is said to have become lost (Ley 1980) because the contingency inherent in this emergence makes it difficult to account for durable traits that might define a self. With non-representational thinking, the subject is seen as derived in practice and is neither singular nor stable (Thrift 1996, 1997; Nash 2000), but is constantly remade through active interactions with other entities, objects and materials. Subjectivity is seen as intensity, multiplicity, productivity and discontinuity (Thrift 1996) rather than embodied and essential, and things outside the human form are seen as constituent factors in human subjectivity. Human consciousness is hereby extended into the non-human world, as the human is seen as co-fabricated with other entities, leading to recognition of more-than-human associations and subjectivities in which agents (entities with the capacity to make a difference, both human and otherwise) are unstable and disembodied (Whatmore 1997, 2006). Ethological or relational perspectives, which consider human beings to reside within rather than stand apart from their environment, consider embodied subjectivity and the world to be reciprocally related (Crowther 1993) in a manner that denies conventional notions of an essential psychological subject contained within a biological human body. Geography has drawn variously upon works such as Whitehead's (1920) ethological concept of nature, Massumi's (2002) consideration of the virtual, and the machinic assemblages of Deleuze and Guattari (Guattari 1995; Deleuze and Guattari 2004), which themselves draw on lines of thought from Bergson, Wittgenstein, Heidegger and Merleau-Ponty, among others. Yet subjectivity remains a vexing issue for non-representational geography, with questions hanging over its existence, nature and relevance (Thrift 2008; Blackman 2010, 2012; Rose 2010; Wylie 2010). Gendlin talks about his philosophy as being a non-representational philosophy of the subject (Gendlin 2001), and as such it offers the potential to invigorate this seemingly intractable disciplinary debate (Banfield 2014).

Gendlin's philosophy is an ethological model in which bodies (not restricted to human bodies) and their environment are one (Gendlin 2001). As in Massumi's work on movement and affect (Massumi 1995, 2002) process comes first, with body and environment being one event. Here, Gendlin gives the example of life going on in the spider's web as much as in its body (Gendlin 2001). The life process makes itself an environment in which it can then go further. Gendlin describes four types of environment: the spectator's environment, or an organism's environment as described by a spectator; the reflexively identical environment, in which body and environment are one event/process; the environment that is arranged by the body-environment event/process; and something within that body-environment that might affect the life process one day, but has not done so yet. Having outlined these four definitions of environment, Gendlin illustrates how they interact and elaborates upon his model to emphasize the role of the implicit in this body-environment process (Banfield 2014).

It is a processual model in which entities emerge from an implicit many through a process that Gendlin calls coordinated differentiation. As outlined earlier, the implicit refers to everything of which we are not explicitly (reflectively) aware. The implicit many recognizes the implicit connectivity between those things that we think about in terms of objects, entities, people, events and so on. While we think of objects and people, for example, as multiple discrete units, they are implicitly connected and they emerge from that implicit connectivity. The implicit (many) is not static but, similar to notions of affect as a sense of compulsion or stimulus in the world (Thrift 2004a), there is an ongoing implying, an indeterminate potential for things to be other than they are. As things happen or change, which Gendlin calls occurring, objects and people become coordinatedly differentiated from the implicit many and from each other, which also changes the indeterminate potential (implying). Consequently, subjects and objects are not essential but are continually renewed through the body-environment process. This process consists, then, of two specific elements, implying and occurring. The implicit is the more-than-logical, and implying is ongoing, playing out through myriad, diverse and inter-locked body-environment processes. The body is not what is in the skin-envelope, but body and environment are interspersed, and the implyings of these multiple body-environment processes "interaffect" each other because they are implicitly connected as part of the implicit many (Gendlin 2001). The continual renewal of the body-environment process within the implicit many means that whatever occurs continues to interaffect the implicit many after it has finished occurring. Similar to Massumi's (2002) virtual swarm, in which participation is prior to participants, Gendlin's process is undivided, because that which is shaped (divided/differenti-ated), is shaped by participating in the shaping (Gendlin 2001), and this shaping is ongoing. Within this, the body's own implying, which Gendlin calls focaling, is akin to Massumi's sensation as the channelling of field potential into local action (Massumi 2002), enabling emergent subjects to enact their own implying as directed or intentional actions (Gendlin 2001).

There is both a distinction and a connection, then, between the implicit, which is a transversal connectedness whereby everything is already interaf-fected by everything else, and someone's individual implicit understand-ing. This individual implicit understanding is connected with the implicit many from which it was coordinatedly differentiated. However, it also has specificity because of its coordinated differentiation, through which it becomes increasingly distinct. The subject that emerges is both connected with the implicit many and unique within it. Just as our formal conceptual systems are (to a large degree) shared or agreed intersubjectively, so we are capable (to a degree) of a shared sense of the implicit through this original interaffecting of everything by everything else.

The capacity within Gendlin's model for a degree of subjective dura-tion, a sense of self, distinguishes it from much non-representational geo-graphical work, which struggles to accommodate human subjectivity and agency. While I deal with this in more detail in Chap. 3, one final point to make is Gendlin's idea of "had space" or "had space-and-time", which

I refer to here as "had" space-time (Gendlin 2001). This is a particular and implicit sense of the connectivity and potentiality within a given body-environment concretion at a particular time, which allows for a personal sense of intersubjectivity and opportunities for action, along with a sense of possible outcomes arising from those actions. This picks up on two types of implying: horizontal implying, in which parts and wholes imply each other through their connectivity within the implicit many; and temporal implying, in which behavioural sequences and outcomes imply each other through their connectivity in the body-environment process of implyings and occurrings (Gendlin 2001). Although the term "had" space is not punctuated in Gendlin's work in the manner that I have chosen to, I do so to make it easier to comprehend the phrase as a Gendlinian term rather than a typographical error.

This implicit sense of a range of behaviour possibilities in a particular situation is far removed from everyday ideas of space as a static surface or container for action. Gendlin's notion of "had" space-time is a more active and processual sensing of space, through which we are implicitly connected to our body-environment process and to other body-environments through their interaffecting (Gendlin 2001). This is an ongoing sensing of how things are and how things might be, consistent with the processuality of connective sensibilities emphasized in non-representational geography (McCormack 2003; Wylie 2010).

Conclusion

Non-representational geography is a diverse field of activity, which I have characterized in terms of a performative turn and a turn to affect. It has been developed more fully theoretically than methodologically, granting it a more robust and distinctive conceptual basis than methodological practice, although both conceptual and methodological challenges remain. This sub-disciplinary domain has been heavily influenced by certain philosophies, particularly those of Spinoza, Massumi, Deleuze and Guattari, which has gifted to non-representational geography certain understandings and vocabularies, which both enable and constrain its thinking and practice. In essence, Gendlin's philosophy is consistent with others already familiar to non-representational geography, although I find Gendlin's emphasis on verbs and processes (e.g. implying, occurring, interaffecting, focaling) more appealing than the planar and machinic terminology that dominates much current non-representational work. I also find Gendlin's understanding of the relation between the implicit and the explicit—granting us the capacity to generate concepts from our pre-reflective understanding—more consistent with non-representational geographical concerns and interests than assumptions of our incapacity to articulate from the affective, which currently prevail.

In several respects, I think that Gendlin offers particular benefits for non-representational geography, bringing alternative perspectives to bear on a range of issues and challenges. Through his writings, Gendlin introduces a kedgeree of new concepts and terms, some of which have an affinity with ideas already familiar in non-representational geography, but his philosophy and methods allow for certain disciplinary debates and challenges to be reworked. I am thinking specifically about the question of human subjectivity and agency, and the uncertainty as to how to work discursively with the pre-reflective. Gendlin's philosophy, vocabulary and psychotherapeutic techniques all offer potential to enrich non-representational geography. My hope, in writing this book, is to initiate and stimulate a thorough engagement with and exploration of this potential. Specifically, in introducing Gendlin to geography I hope to:

1. Contribute a philosophical counterweight to the dominance of Spinozan–Massumian–Deleuzian influences on non-representational work, with potential for the adoption and development of new understandings and conceptual vocabularies.
2. Encourage and facilitate, through this philosophy, further rethinking of subjectivity, intersubjectivity and agency.
3. Stimulate reconsideration of our capacity to access and articulate (explicate) from the affective by introducing a philosophy that bridges seeming distinctions between emotional and affectual geographies.
4. Initiate the adoption and development of particular explicatory techniques, characterized as an intimate form of knowledge, through which we might both generate new means of working with the implicit (affective) and develop more affectively attuned disciplinary vocabularies.
5. Prompt a reconfiguration of what it is to articulate from affective or implicit understandings, not in terms of reduction or capture, but in more affirmative terms, such as creating words that do count and that keep alive the creative potential of affect.

In short, I hope to introduce, explore, illustrate and elaborate a philosophy and associated methodology which has the potential to fend off criticisms of non-representational geography as misleading, incoherent and hypocritical. In the next chapter, I detail the research through which I engaged with Gendlin's work and experimented with his explicatory techniques, situated within the geographies of artistic practice.

REFERENCES

Anderson, Ben 2006 Becoming and being hopeful: Towards a theory of affect. *Environment and Planning D: Society and Space* 24 733–752

Anderson, Ben 2013 Affect and emotion. *in:* Johnson, NC, Schein, RH & Winders, J (eds.) *The Wiley-Blackwell companion to cultural geography.* Wiley, Hoboken 452–462

Anderson, Ben & Harrison, Paul 2006 Commentary: Questioning affect and emotion. *Area* 38 333–335

Anderson, Ben & Harrison, Paul 2010a The promise of non-representational theories. *in:* Anderson, B & Harrison, P (eds.) *Taking-place: non-representational theories and geography.* Ashgate, Farnham 1–34

Anderson, Ben & Harrison, Paul 2010b *Taking-place: non-representational theories and geography.* Ashgate, Farnham

Banfield, Janet 2014 *Towards a non-representational geography of artistic practice.* Unpublished doctoral thesis, University of Oxford, Forthcoming online: https://ora.ox.ac.uk:443/objects/uuid:dd12e1c4-f222-435b-adc0-c1bb68e4f4ac

Bertelsen, Lone & Murphie, Andrew 2009 An ethics of everyday infinities and powers: Felix Guattari on affect and the refrain. *in:* Gregg, M & Seigworth, GJ (eds.) *The affect theory reader.* Duke University Press, Durham, NC; London 138–157

Bissell, David 2010 Placing affective relations: Uncertain geographies of pain. *in:* Anderson, B & Harrison, P (eds.) *Taking-place: non-representational theories and geography.* Ashgate, Farnham 79–97

Blackman, Lisa 2010 Embodying affect: Voice-hearing, telepathy, suggestion and modelling the non-conscious. *Body and Society* 16 163–192

Blackman, Lisa 2012 *Immaterial bodies: affect, embodiment, mediation.* SAGE, London

Blackman, Lisa & Venn, Couze 2010 Affect. *Body and Society* 16 7–28

Bondi, Liz 2005 Making connections and thinking through emotions: Between geography and psychotherapy. *Transactions of the Institute of British Geographers* 30 433–448

Buttimer, Anne 1976 Grasping the dynamism of lifeworld. *Annals of the Association of American Geographers* 66 277–292

Clough, Patricia T 2010 Afterword: The future of affect studies. *Body and Society* 16 222–230

Crang, Mike 1998 *Cultural geography* Routledge, London

20 J. BANFIELD

Crowther, Paul 1993 *Art and embodiment from aesthetics to self-consciousness* Oxford University Press, Oxford
Curti, Giorgio H, Aitken, Stuart C, Bosco, Fernando J & Goerisch, Denise D 2011 For not limiting emotional and affectual geographies: a collective critique of Steve Pile's 'Emotions and affect in recent human geography'. *Transactions of the Institute of British Geographers* 36 590–594
Daniels, Stephen 1985 Arguments for a humanistic geography. *in:* Johnston, R J (ed.) *The future of geography.* Methuen and Co. Ltd, London 143–158
Daniels, Stephen 2004 Marxism, culture, and the duplicity of landscape. *in:* Thrift, N & Whatmore, S (eds.) *Cultural geography: critical concepts in the social sciences, vol 2 practising culture.* Routledge, London 17–44
Deleuze, Gilles & Guattari, Félix 2004 *A thousand plateaus: capitalism and schizophrenia.* Continuum, London
Dewsbury, John-David 2009 Affect. *in:* Thrift, N & Kitchin, R (eds.) *International encyclopedia of human geography.* Elsevier Science, Amsterdam; London 20–24
Dewsbury, John-David 2010a Language and the event: the unthought of appearing worlds. *in:* Anderson, B & Harrison, P (eds.) *Taking-place: non-representational theories and geography.* Ashgate, Farnham 147–160
Dewsbury, John-David 2010b Performative, non-representational, and affect-based research. *in:* Delyser, D, Herbert, S, Aitken, S, Crang, M & Mcdowell, L (eds.) *The SAGE handbook of qualitative geography.* SAGE, Los Angeles; London 321–334
Dewsbury, John-David & Naylor, Simon 2002 Practising geographical knowledge: Fields, bodies and dissemination. *Area* 34 253–260
Dirksmeier, Peter & Helbrecht, Ilse 2008 Time, non-representational theory and the "performative turn" – Towards a new methodology in qualitative social research. *Forum Qualitative Social Research* 9 Art 55
Dixon, Deborah & Straughan, Elizabeth R 2013 Affect. *in:* Johnson, NC, Schein, RH & Winders, J (eds.) *The Wiley-Blackwell companion to cultural geography.* Wiley, Hoboken 36–38
Driver, Felix 2003 On geography as a visual discipline. *Antipode* 35 227–231
Featherstone, Mike 2010 Body, image and affect in consumer culture. *Body and Society* 16 193–221
Game, Ann 2001 Riding: Embodying the centaur. *Body and Society* 7 1–12
Gendlin, Eugene T 1980 Imagery is more powerful with focusing: theory and practice. *in:* Shorr, JE, Sobel, GE, Robin, P & Connella, JA (eds.) *Imagery. Its many dimensions and applications.* Plenum Press, New York; London 65–73
Gendlin, Eugene T 1989 Phenomenology as non-logical steps. *in:* Kaelin, EF & Schrag, CO (eds.) *American Phenomenology: origins and developments.* Kluwer, Dordrecht 404–410

Gendlin, Eugene T 1993 Words can say how they work. *in:* Crease, RP (ed.) Proceedings, Heidegger Conference. Stony Brook, State University of New York

Gendlin, Eugene T 1995 Crossing and dipping: Some terms for approaching the interface between natural understanding and logical formulation. *Minds and Machines* 5 547–560

Gendlin, Eugene T 1997 The responsive order: A new empiricism. *Man and World* 30 383–411

Gendlin, Eugene T 2001 *A Process Model* The Focusing Institute, New York

Gendlin, Eugene 2009a What first and third person processes really are. *Journal of Consciousness Studies* 16 332–362

Gendlin, Eugene T 2009b We can think with the implicit, as well as with fully-formed concepts. *in:* Leidlmair, K (ed.) *After cognitivism: a reassessment of cognitive science and philosophy.* Springer, London, New York 147–161

Gibbs, Anna 2009 After affect: sympathy, synchrony and mimetic communication. *in:* Gregg, M & Seigworth, GJ (eds.) *The affect theory reader.* Duke University Press, Durham, NC; London 186–205

Greenhough, Beth 2010 Vitalist geographies: life and the more-than-human. *in:* Anderson, B & Harrison, P (eds.) *Taking-place: non-representational theories and geography.* Ashgate, Farnham 37–54

Grossberg, Lawrence 2009 Affect's future: Rediscovering the virtual in the actual. *in:* Gregg, M & Seigworth, G J (eds.) *The affect theory reader.* Duke University Press, Durham, NC; London 309–338

Guattari, Félix 1995 *Chaosmosis: an ethico-aesthetic paradigm.* Power Publications, Sydney

Ingold, Tim 2011b *The perception of the environment: essays on livelihood, dwelling and skill* Routledge, London

Ley, David 1980 *Geography without man: a humanistic critique* School of Geography, University of Oxford, Oxford

Leys, Ruth 2011 The turn to affect: A critique. *Critical Inquiry* 37 434–472

Livingstone, David N 1992 *The geographical tradition: episodes in the history of a contested enterprise* Blackwell, Oxford

Lorimer, Hayden 2005 Cultural geography: The busyness of being 'more-than-representational'. *Progress in Human Geography* 29 83–94

Lorimer, Hayden 2007 Cultural geography: Wordly shapes, differently arranged. *Progress in Human Geography* 31 89–100

Lorimer, Jamie 2010 Moving image methodologies for more-than-human geographies. *Cultural Geographies* 17 237–258

Massumi, Brian 1995 The autonomy of affect. *Cultural Critique* 83–109

Massumi, Brian 2002 *Parables for the virtual: movement, affect, sensation* Duke University Press, Durham, NC; London

McCormack, Derek P 2003 An event of geographical ethics in spaces of affect. *Transactions of the Institute of British Geographers* 28 488–507

McCormack, Derek P 2004 Introduction: techniques and non-representation. *in:* Thrift, N & Whatmore, S (eds.) *Cultural geography: critical concepts in the social sciences, vol 2 practising culture.* Routledge, London 3–16

McCormack, Derek P 2005 Diagramming practice and performance. *Environment & Planning D: Society & Space* 23 119–147

McCormack, Derek 2006 For the love of pipes and cables: A response to Deborah Thien. *Area* 38 330–332

Merleau-Ponty, Maurice 1995 *Phenomenology of perception* Routledge, London

Nash, Catherine 2000 Performativity in practice: Some recent work in cultural geography. *Progress in Human Geography* 24 653–664

Oakes, Tim & Price, Patricia L 2008 *The cultural geography reader* Routledge, London

Parry, Joseph D & Wrathall, Mark 2011 Introduction. *in:* Parry, JD (ed.) *Art and phenomenology.* Routledge, Abingdon 1–8

Pels, Dick, Hetherington, Kevin & Vandenberghe, FrÈdÈric 2002 The status of the object – performances, mediations, and techniques. *Theory Culture & Society* 19 1–20

Pile, Steve 2010 Emotions and affect in recent human geography. *Transactions of the Institute of British Geographers* 35 5–20

Rose, Gillian 1993 *Feminism and geography: the limits of geographical knowledge.* Polity Press, Cambridge

Rose, Mitch 2010 Envisioning the future: ontology, time and the politics of non-representation. *in:* Anderson, B & Harrison, P (eds.) *Taking-place: non-representational theories and geography.* Ashgate, Farnham 341–361

Samuels, Marwyn S 1978 Existentialism and human geography. *in:* Ley, D & Samuels, MS (eds.) *Humanistic geography: prospects and problems.* Croom Helm, London 22–40

Simpson, Paul 2009 'Falling on deaf ears': a postphenomenology of sonorous presence. *Environment and Planning A* 41 2556–2575

Thien, Deborah 2005 After or beyond feeling? A consideration of affect and emotion in geography. *Area* 37 450–454

Thrift, Nigel J 1996 *Spatial formations* SAGE, London

Thrift, Nigel J 1997 The still point. *in:* Pile, S & Keith, M (eds.) *Geographies of resistance.* Routledge, London 124–151

Thrift, Nigel 2004a Intensities of feeling: Towards a spatial politics of affect. *Geografiska Annaler, Series B: Human Geography* 86 57–78

Thrift, Nigel 2004b Summoning life. *in:* Thrift, N & Whatmore, S (eds.) *Cultural geography: critical concepts in the social sciences, vol 2 practising culture.* Routledge, London 431–455

Thrift, Nigel J 2008 *Non-representational theory: space, politics, affect* Routledge, London; New York

Thrift, Nigel J; Harrison, Paul & Anderson, Ben 2010 "The 27th letter": an interview with Nigel Thrift. *in:* Anderson, B & Harrison, A (eds.) *Taking-place: non-representational theories and geography.* Ashgate, Farnham 183–198

Tuan, Yi-Fu 1971 Geography, phenomenology, and the study of human nature. *Canadian Geographer-Geographe Canadien* 15 181–193

Wetherell, Margaret 2012 *Affect and emotion: a new social science understanding* SAGE, London

Whatmore, Sarah 1997 Dissecting the autonomous self: Hybrid cartographies for a relational ethics. *Environment and Planning D: Society & Space* 15 37–53

Whatmore, Sarah 2006 Materialist returns: practising cultural geography in and for a more-than-human world. *Cultural Geographies* 13 600–609

Whitehead, Alfred N 1920 *The concept of nature: Tarner lectures delivered in Trinity College, November, 1919* Cambridge University Press, Cambridge

Wylie, John 2005 A single day's walking: Narrating self and landscape on the South West Coast Path. *Transactions of the Institute of British Geographers* 30 234–247

Wylie, John 2006 Depths and folds: on landscape and the gazing subject. *Environment and Planning D: Society & Space* 24 519–535

Wylie, John 2010 Non-representational subjects? *in:* Anderson, B & Harrison, P (eds.) *Taking-place: non-representational theories and geography.* Ashgate, Farnham 99–114

CHAPTER 2

Geographies of Artistic Practice

Abstract This chapter describes the development and convergence of research methods in geography. It details Banfield's own research design, explores the growth in the use of video- and practice-based methods—in which researchers actively participate in the practice that they are researching—and psychologically informed methods—which use psychotherapeutic techniques for non-clinical purposes. Banfield proposes that these developments reflect an increasing interest across the social sciences in pre-reflective and embodied experience, consistent with the rise of non-representational thinking, even if it is not explicitly informed by such thinking. The chapter describes geographical research into artistic spatial experiences through which Eugene Gendlin's potential contribution to geography is explored.

INTRODUCTION

Whereas the previous chapter focused on non-representational geography as a broad domain of academic activity with particular thematic interests, conceptual challenges and methodological needs, and introduced Eugene Gendlin to the discipline, in this chapter I situate my research more specifically within geographies of art. I begin by briefly charting recent developments in geographical engagements with art, and highlighting the increasingly practice-based and collaborative nature of this work, which

© The Author(s) 2016
J. Banfield, *Geography Meets Gendlin*,
DOI 10.1057/978-1-137-60440-8_2

is paralleled by methodological developments across the social sciences. I characterize these developments as a progressive coming together of academic concerns for the practical and the sensual, with an increasing interest in visual methods. Against this interdisciplinary background I situate my specific research aims, design and methods to lay the groundwork for the empirical material that is presented in Part 2 (in relation to Gendlin's theoretical ideas) and Part 3 (in relation to Gendlin's methodological implications).

INTERDISCIPLINARY INTEGRATION

With its close associations with drawing, mapping and diagrams, geography is often characterized as a visual discipline, and this generic characterization has been proposed as a possible reason for a lack of an identifiable practice-led sub-discipline which is concerned with the visual (Driver 1995, 2003; Rose 2003; Thornes 2004; Tolia-Kelly 2012; Jacobs 2013). With a tighter focus on geographical engagement with formal artworks, the past few decades have seen the progression of geography's understandings of art from a descriptive practice, through interpretative and formative practices, to performative and transformative understandings. The latter two in particular, have stimulated the development of a diffuse field of activity involving a range of artistic media, geographical themes and artistic doings (Hawkins 2013, 2015; Madge 2014; Banfield 2016a).

In contrast to earlier geographic use of artistic practice primarily as a means of description, as in the case of gazetteers of colonial explorers or early twentieth-century analyses of landscape (Cornish 1935; Marston and de Leeuw 2013), the geographical analysis of landscape paintings by writers such as Dennis Cosgrove and Stephen Daniels moved geography's use of art from description to the interpretation of both wider social conditions and the particular social status of the artist or their patron (Cosgrove 1984, 1985; Daniels 1984; Daniels and Cosgrove 1988; Prince 1988; Marston and de Leeuw 2013; Banfield 2014).

A more active geographical understanding of art recognizes the power of art to shape as well as to reflect socio-spatial environments, in a productive or formative understanding of art and artistic practice. This idea is most clearly seen in monumental or installation art, in which the art installed changes the landscape into which it is installed (Morris and Cant 2006; Morris 2011), for example in the case of statues (e.g. The Statue of Liberty, The Angel of the North) and land art (e.g. Mount Rushmore, the Cerne Abbas Giant). However, it can also be seen in the preservation within landscape paintings of landscape ideals, becoming revered as meaningful symbols of regionally distinctive cultural landscapes which then take on totemic value, as in the case of Constable Country (Osborne 1988). In a similar vein, critical social geographers have also highlighted the productive nature of art, for example in relation to the contested nature of home-based studio spaces, whereby the studio as a work environment must negotiate with other demands in the domestic setting. Equally, on a community level, the role of artists in urban regeneration has been emphasized, whether through the location of studios to create an arts quarter, through the production of public art as a means of public engagement, or through culturally oriented development strategies to encourage an arts-based economy (Deutsche 1988; Zukin 1989, 1995; Bain 2006, 2007). Across the varied artistic products and practices considered, such studies simultaneously portray artistic and social practices as shapers of landscapes, and landscapes as shapers of artistic and social practices (Banfield 2014, 2016a).

Art has also been studied as a personal form of spatial knowledge-making. In such studies, artworks are informative not solely in terms of the social life they depict but in terms of the personality and values of the artist behind the work. Examples here include David Crouch and Mark Toogood's consideration of Peter Lanyon's strong identity as Cornish, and David Matless and George Revill's account of Andy Goldsworthy's erratics: land sculptures which work the land in a practical engagement with the rhythms of a rural life with which he is familiar (Matless and Revill 1995; Crouch and Toogood 1999; Banfield 2014, 2016a). Here, expression comes to the fore as personal qualities are performed and made accessible through the artwork. This performative understanding of art does not preclude any of the previous understandings, as the finished works can still be conceptualized in descriptive terms, they are still capable of interpretation, and their association with specific geographic locations might still function in a formative manner. However, this performative understanding of art emphasizes more emergent senses of the subjectivities and spatialities they generate, as the artist is considered to be influenced by the art-making process.

Rather than relying on detached observation and the visual communication of representational content, some approaches to spatial knowledge-making employ artistic practice as an alternative means of accessing and experiencing the environment at hand, and of communicating that embodied and affective experience. Examples here, which often draw on ideas of dwelling or practical engagement in the landscape (Wylie 2011), include individual and reflexive accounts of the writer's own aesthetic and artistic engagements with places (Edensor 2000; Wylie 2005). The idea of art as a practice of dwelling connects strongly with a final understanding of art as transformative, in both a personal sense by facilitating self-understanding and generating a sense of wellbeing, and in a social sense by providing a means for subaltern voices and experiences to be heard through social or participatory activism (Parr 2006, 2007; Tolia-Kelly 2007). Key to these transformative understandings is a belief that art and artistic practice can in some way grant access to unspoken or inarticulable aspects of experience, or that it allows the expression of sensations and emotions more powerfully than words will allow, somehow connecting more immediately and intimately both with our own inner experience and intersubjectively with others (Bondi 2005; O'Neill 2008; Hogan 2009; Pink et al. 2011; Hogan and Pink 2012).

As a means of experiencing the environment and as a practice of dwelling within it, artistic practice is no longer confined to the artist, but can be employed by artists and non-artists alike as a means of experience and expression. This democratization of artistic practice as spatial knowledge-making opens up the possibility of myriad new approaches to geographies of art in which artist, artwork, place relations, artistic practice and research practice can all be co-located. Despite a current lack of consensus concerning the degree of artistic proficiency required on the part of the geographer wishing to undertake artistic practice within their research (Lafrenière and Cox 2013; Marston and de Leeuw 2013; Banfield 2014, 2016a; Madge 2014; Hawkins 2015), the recent burgeoning of collaborative and practice-based approaches to such inquiries is testament to their potential.

Further strengthening this democratization of artistic practice in research is the connection between this transformative understanding of art and the use of psychotherapeutic methods for non-therapeutic purposes, using such methods as an instrument of knowledge rather than a clinical tool (Frosh, in Oliver 2003). While not denying the considerable ethical issues surrounding the pursuit of therapeutic objectives by a researcher not trained in such treatments, or those surrounding the involvement of individuals with mental health difficulties in research without sufficient protections and provisions, such caveats need not preclude the use of therapeutic methods for non-therapeutic purposes with non-clinical and inexpert participants. If arts-based psychotherapeutic methods are used in clinical settings with non-artists to increase the articulation of deep-seated experiences, the question arises as to why research into the articulation of such experiences should not use similar methods for non-therapeutic aims.

Since the 1990s, the use of psychotherapeutic epistemologies and methods within geography has grown substantially, in particular those of psychodynamic and humanistic flavours (Bondi 1999, 2005; Philo and Parr 2003), and has been linked to developments in qualitative and arts-based methods (Bondi 1999, 2003; Parr 2007). Both Hester Parr and Liz Bondi draw on their involvement with people with mental health difficulties in their academic geographical work. Most notably for current purposes, Parr talks about using film-making in fieldwork as enabling different participative and social outcomes, and observes that participants talk about their involvement in her fieldwork as therapeutic, even though their artistic activity in fieldwork is not intended as a clinical practice. In addition, although commenting that there is arguably a special relation between people with mental health difficulties and visual imagery, which might cast doubt on the appropriateness or effectiveness of employing arts-based psychotherapeutic methods with a non-clinical population, the arguable nature of this special relation encourages further exploration in geographical research (Parr 2006, 2007).

In summary, geography's understanding of, and interest in, art have developed from a primarily descriptive tool through an analytical target, and then through a constitutive force, both socially and individually, and latterly to a performative and transformative medium. These developments exemplify a trend towards increasingly practice-based and collaborative approaches, which also draw on increasing disciplinary engagement with psychotherapeutic methods, including those that employ their own practice-based artistic activities as a means of accessing and articulating deeply felt experiences that are difficult to communicate. While geographical interest in art as a subject of inquiry has not waned, the growth of a broader disciplinary concern for practices has also led to an explosion of interest in art as a method of geographical inquiry.

Geographical research into art frequently includes the creation and exhibition of artworks within a research process, such as installation, monumental and environmental art forms, as well as drawn or painted pictures (Morris and Cant 2006; Morris 2011). Increasingly, individual researchers are undertaking their own creative practice to communicate their research experiences creatively, while others conduct their research in collaboration with practising artists, producing either one-off projects or sustained programmes of interaction. Harriet Hawkins, for example, distinguishes between those geography–art relationships whereby geographers analyse artworks (dialogues) and those in which geographers become practitioners or collaborators (doings), and has engaged in collaborative activity through an artist-led community arts project (Hawkins 2011, 2015). Other artist–geographer collaborations include an exploration of the means by which artworks conjure narratives of place that challenge dominant narratives of globalization, and an account of working together to develop common interests between artistic and geographic practices (Mackenzie and Taylor 2006, Foster and Lorimer 2007). Artistic methods have been employed as a means to examine both art itself, as in Hawkins' (2015) collaborative work with Annie Lovejoy, and broader geographical concerns, as in Tolia-Kelly's work exploring post-migration landscape relations in the Lake District, typically considered an icon of Englishness (Tolia-Kelly 2007). This trend towards the use of art as a research method is not unique to geography, but can be seen across the social sciences in a coming together of qualitative, collaborative, arts-based and practice-based methods (Banfield 2016a).

My specific concern is with the emphasis on the reputed power of the visual to access and express something of our ineffable inner experience, with increasingly practice-based and often increasingly ethnographically informed approaches to visual research. Alongside resurgent interest in the visual, and a transformation in the understanding of the visual as multi-sensory, social science research is witnessing a transformation in the tools at its disposal for visual research, as evidenced in the growth of film/video methods and the availability of digital and online formats (Jacobs 2013). Collectively, these developments are proposed to offer routes to new forms of knowledge through increasing integration of arts-based and ethnographic methods (Hogan and Pink 2012; Knoblauch 2012; Pauwels 2012; Pink 2012; Jacobs 2013). Video has been used in ethnographic practice since the 1980s, and although it is becoming increasingly popular as a research tool, especially for real-time data collection in practice-based inquiry (Pink 2011b; Jacobs 2015; Pink et al. 2015), moving imagery is still used less frequently than static imagery in qualitative research (Banks 2007).

A distinction can be drawn here between video, broadly conceived as a real-time and relatively objective record of research activity, and film-making, often considered to be more collaborative, subjective and creative (Banks 2007; Cranny-Francis 2009; Jacobs 2015; Pink et al. 2015). While video is increasingly used as a research record, film-making remains uncommon, and video/film-based research is still sufficiently under-developed that it has been characterized as lacking a unified body of methods (Knoblauch 2012), perhaps further limiting its uptake. Within geography specifically, few researchers use video for data collection and almost none engage with experimental film-making, meaning that the discipline is yet to realize the full potential of video as a methodology (Garrett 2010). The transformation in the understanding of visual media and its re-situation as multisensory (Gibbs 2009; Blackman 2010, 2012; Featherstone 2010) look set to stimulate further engagement with video methods in the context of growing efforts to study non-visual aspects of perception and to explore affective experience beyond conventional methods (Pink 2009, 2011a, b, c, 2012; Merchant 2011; Pink et al. 2011; Hogan and Pink 2012; Pauwels 2012).

The use of imagery, and especially video, is also increasingly being used as stimulus material in elicitation methods, whereby reviewing an image or video provides an opportunity to revisit the practice through which it was produced. Such video-elicitation methods are reported: to allow the discovery or identification of additional features of significance in the practice that had not been evident during the real-time experience; to stimulate unexpected reviewer responses; and to allow the examination of the past as a former present by reconstructing one's own experience (Garrett 2010; Knoblauch 2012; Laurier and Philo 2012; Pink et al. 2015). Questions and debates persist concerning the efficacy of video-elicitation on the grounds that video is as situated and constructed as any other data, or the grounds that the commentary elicited during review will be influenced by the circumstances of that viewing (Banks 2007; Pink 2011a; Mondada 2012). However, the capacity for video to enable the revisiting of previous practical and embodied experience, either through the reconstruction of one's own experience or through the imagination of the experiences of others, offers the potential to open up new avenues for inquiry into embodied, sensory and affective elements of experience (Merchant 2011). Consequently, with its emphasis on implicit or affective aspects of experience, non-representational geography's identified need for methodological innovation and invigoration (Blackman and Venn 2010; Clough 2010; Lorimer 2010; Pile 2010) could benefit considerably from the purported capacity for both arts-based and video-based research methods to facilitate access to these ineffable qualities.

In all, these developments indicate a coming together of arts-based, video-based and practice-based methods across the social sciences, which are increasingly interested in the visual, the practical and the sensual. It is among these interdisciplinary developments and their methodological implications that my own research nestled. This research, informed by a non-representational attitude, explored the emergence of spatialities and subjectivities from within artistic practices, and used both practice-based and video-elicitation methods, in combination with interview techniques inspired by Gendlin's explicatory methods, in an attempt to access pre-reflective aspects of these practices.

RESEARCH METHOD AND PARTICIPANTS

In the spirit of a non-representational understanding of subjectivity and spatiality, which recognizes the emergence of self and place through each other in practice (Thrift 1996, 1997; Doel 1999, 2000; Nash 2000; Anderson and Tolia-Kelly 2004; Massey 2005; Wylie 2005, 2006), my research explored subjectivity and spatiality in their artistic unfolding. Disciplinary debates continue as to whether research that generates images should be described as visual culture or image-making rather than art, and over the need for those engaged in such research to be proficient or trained in art to be able to refer to the images generated as art (Tolia-Kelly 2012; Lafrenière and Cox 2013; Marston and de Leeuw 2013; Madge 2014; Banfield 2016a; Hawkins 2015). However, I use the term artistic practice and refer to participants as artists, as they all engaged in image-making in a formal capacity.

My research (Banfield 2014, 2016a) involved semi-structured interviews; two production sessions in which the researcher (a hobby artist) and artist/s worked alongside each other on their respective artworks; and stimulated recall and review of the research activities based on audio-visual recordings of the production sessions. This enabled me to generate both retrospective consideration and communication, and real-time articulation, of the generation of spatiality and subjectivity in these artists' practices.

Potential participants were identified through the promotional materials associated with a local annual arts festival held in Oxfordshire, UK. Criteria for consideration included: the need for participants' practices to involve media and materials that could reasonably be transported for use in the production sessions; my need to maximize the number of exhibition venues that I could visit during the festival (thereby focusing on local clusters of exhibitions); the need to identify individual artists with their specific medium of choice (thereby excluding group exhibitors); and the need for participants to be currently practising their art (thereby excluding anybody exhibiting historic work only). I worked with twelve participants over a six-month period during 2012, most of whom were professional artists and had attained formal art qualifications, although two were hobbyists. All professionally practising participants had taken up their artistic career following previous employment. All participants were female, and only one participant was under the age of 40 (average age 58). These demographic characteristics were not intentional, but were a function of the criteria that I adopted and the preponderance of particular media among different demographic groups within the promotional brochure for the festival (Banfield 2014, 2016a). For this publication, I focus on work undertaken with four participants, who were particularly informative and detailed with regard both to the processes by which subjectivity and spatiality come into being through their practices, and to pertinent aspects of Gendlin's work, to provide as succinct and vivid an account as possible of Gendlin's relevance to geography as identified through my research.

Two of these participants took part on an individual basis throughout the research. Jane (age 71) has retired from a career in art education, during which she taught in numerous countries, and uses a variety of media in her practice, which often depicts landscapes or scenery. Laura (age 36) undertook degrees in fine art after an initial career in publishing, and is establishing her art career in Oxford. Laura's medium of choice is oil paint and oil pastel, and she describes her work as being semi-abstract and about her experience of the spirit of place (Banfield 2014, 2016a).

The other two participants on whose accounts and practices I draw here conducted their preliminary and closing interviews individually, but took part in the production sessions as a group (with another research participant), because they routinely work together on a project focused on capturing dance in painted form alongside their own practices. Susan (age 57) has completed art courses in both South Africa and the UK and worked as an illustrator before committing to her own practice approximately twenty years ago. Susan works in mixed media, such as print, paint and jewellery, and says that her work is inspired by her life experiences. Clare (age 58) initially worked as a film editor and researcher before embarking on her artistic career. Clare works with a range of media including oils, printmaking and watercolour, and much of her work involves prints of monumental figures in unconventional positions that invite an explanation (Banfield 2014, 2016a).

As for me, I was 36 at the time of the research and I have no formal artistic training, although I have maintained my hobby practice for over twenty years, since reluctantly dropping the subject at school at the age of 16. Prior to undertaking the research, my medium was hand embroidery, in which the works I designed and created were primarily landscapes and three-dimensional pieces, typically butterflies or flowers. Although not a professionally practising artist, I have participated in the arts festival through which I identified potential participants and previously designed and maintained a website promoting my practice. Since undertaking the research, which brought the opportunity to work with different artistic materials, my practice has broadened beyond embroidery, as I have re-engaged with painting (mainly watercolour) and I sometimes work with pastels.

The opening interviews were semi-structured and held at the participant's own home or studio, depending on their circumstances and preferences. These interviews explored the participants' artistic practices, development and qualifications, their media and subjects of choice, and their experiences of their artistic practice over time.

Production sessions were intended to provide a real-time account of participant practices to supplement the interview account of their engagement with art. I hoped to elicit information from participants while doing their artwork about aspects of their artistic practices, which might not be articulated in a conventional interview setting (Banfield 2014, 2016a). My own engagement with artistic practice during these sessions was intended to provide a basis for comparison between our respective practices. By providing conditions in which participants could notice similarities and differences between our respective practices, I sought to make participants more explicitly aware of aspects of their own practice that would normally remain below their conscious awareness, for example because it has become habituated. To maximize the effectiveness of this comparative approach I used, wherever possible, materials that contrasted strongly with those used by the participant/s, for example by working in a dry medium when the participant worked in a wet medium.

In the production sessions, I also wanted to push participants out of the comfort zone of their customary practice in a further attempt to make them more explicitly aware of that customary practice. I did this by varying the materials with which participants worked and by changing where we held the production session. For example, a participant who usually worked in oil paints might have used these materials in the first production session, then switched to watercolours or pastels in the second. Similarly, the first production session might have been held at a participant's studio, where they would normally do their practice, but for the second production session we might go off-site, painting in situ. By changing the conditions of participants' practices so that something was out of the ordinary for them, I aimed to provide an additional means of comparison to elicit more information from them about their customary practices (Banfield 2014, 2016a, 2016b).

Closing interviews explored participants' experiences of, and reflections about, the research process, they followed up on specific points from the previous sessions which seemed important or interesting (for example, if a participant had articulated any discoveries about their practice), and they involved video-elicitation techniques. While the elicitation techniques used, which constitute the bulk of my methodological engagement with Gendlin's work, are specified in detail in Part Three, I outline here a brief summary of my analytical approach to the verbal and visual data that the interviews and production sessions generated.

The interview and production session transcripts were analysed using descriptive phenomenological analysis (Giorgi 1992, 2009). In brief, this method stipulates that researchers should read through a transcript in its entirety first, to gain a sense of the whole, then re-read the transcript to identify units of meaning, and rephrase these meaning units into psychologically relevant expressions, and finally seek unified variations in these expressions to identify the essential features of an experience or phenomenon. While I do not dwell here on the detail of my engagement with phenomenological research methods (for details, see Banfield 2016b), it is worth highlighting that I was not looking for a psychological account, nor for essential features or structures of these artistic experiences, as non-representational geography's post-phenomenological emphasis on emergence would have been at odds with such an approach. I did, however, analyse the data for consistencies and discrepancies between interview and practice-based accounts, between customary and atypical practices, and between participants, to generate as full an impression as possible of participants' artistic practices.

For the visual recordings of the production sessions, I sought to develop an analytical approach that mirrored the one I used for the transcripts. Consequently, I viewed the footage the first time in its entirety to gain a sense of the whole and then viewed it again to identify units of meaning in terms of, for example, the nature, speed and direction of movement, changes in posture, material and movement, and any incidental events or occurrences, such as a sudden breeze or a bird flying past. This enabled me to generate a sense of actions and dispositions that seemed to be typical and atypical within each production session. I then identified and extracted from the footage a selection of short clips, capturing a mixture of both typical and atypical behaviours, for use in the video-elicitation part of the closing interview. During these closing interviews, I asked participants to watch a number of extracted clips and respond to what they were seeing with regard to their actions, intentions and experiences at the time (Banfield 2014). Different Gendlin-inspired elicitation techniques were employed with different participants, which are addressed in more detail in Part Three.

While this section has provided an overview of my research participants, aims and methods, more detail is available elsewhere with specific reference to: the combination of interview and practice-based research methods (Banfield 2016a); the applicability of descriptive phenomenological analysis to practice-based methods (Banfield 2016b); and the research overall, including issues of data quality, reflexivity and ethics (Banfield 2014).

Conclusion

While geography's engagement with art has acquired an increasingly practical flavour, artistic and other visual practices have grown in popularity across the social sciences, and academic concern with practices and embodied experience has become established in a number of disciplines. This coming together of research interests and methods is consistent in character with a non-representational focus on matters of practice and sensibility, a convergence of philosophical and methodological emphases, which is reflected in my own research into the emergence of spatiality and subjectivity in artistic practice.

My research employed both practice-based methods and visually stimulated recall techniques in order to explore the practical embodied experience of artistic doing. As such, the methodology responded to an identified under-representation in the disciplinary literature of methods concerned with non-cognitive embodied experience, and of geographers practising or doing the visual in research that considers visual or artistic practice not just as a form of record but a site of knowing and thinking (Crang 2002, 2003; Thornes 2004; Tolia-Kelly 2012; Hawkins 2015). It also employed two means of enhancing access to pre-reflective aspects of experience: comparative and atypical practices; and Gendlin-informed interview techniques. Gendlin's techniques incorporate both linguistic and artistic elements, essentially using representational practices to access the non-representational, and potentially contributing to philosophical and methodological debates in non-representational geography concerning the relation between the explicit (representational) and implicit (non-representational) and our capacity to access and apprehend the latter. The meaning, relevance and potential within Gendlin's philosophical work forms the focus of Part 2; the nature, application and implications of Gendlin's methodological work forms the focus of Part 3.

References

Anderson, Ben & Tolia-Kelly, Divya 2004 Matter(s) in social and cultural geography. *Geoforum* 35 669–674

Bain, Alison L 2006 Resisting the creation of forgotten places: artistic production in Toronto neighbourhoods. *Canadian Geographer* 50 417–431

Bain, Alison L 2007 Claiming and controlling space: Combining heterosexual fatherhood with artistic practice. *Gender Place and Culture* 14 249–265

Banfield, Janet 2014 *Towards a non-representational geography of artistic practice.* Unpublished doctoral thesis, University of Oxford, Forthcoming online: https://ora.ox.ac.uk:443/objects/uuid:dd12e1c4-f222-435b-adc0-c1bb68e4f4ac

Banfield, Janet 2016a Knowing between: generating boundary understanding through discordant situations in geographic-artistic research. *Cultural Geographies*, 23 459–473

Banfield, Janet 2016b Descriptive phenomenological analysis: practical innovations in geographies of artistic practice. *SAGE Research Methods Cases*, Online resource: http://dx.doi.org/10.4135/9781446273050155953361

Banks, Marcus 2007 *Using visual data in qualitative research.* SAGE, Los Angeles; London

Blackman, Lisa 2010 Embodying affect: Voice-hearing, telepathy, suggestion and modelling the non-conscious. *Body and Society* 16 163–192

Blackman, Lisa 2012 *Immaterial bodies: affect, embodiment, mediation.* SAGE, London

Blackman, Lisa & Venn, Couze 2010 Affect. *Body and Society* 16 7–28

Bondi, Liz 1999 Stages on journeys: some remarks about human geography and psychotherapeutic practice. *Professional Geographer* 51 11–24

Bondi, Liz 2003 Meaning-making and its framings: a response to Stuart Oliver. *Social & Cultural Geography* 4 323–327

Bondi, Liz 2005 Making connections and thinking through emotions: Between geography and psychotherapy. *Transactions of the Institute of British Geographers* 30 433–448

Clough, Patricia T 2010 Afterword: The future of affect studies. *Body and Society* 16 222–230

Cornish, Vaughan 1935 *Scenery and the sense of sight* The University Press, Cambridge, England

Cosgrove, Dennis E 1984 *Social formation and symbolic landscape* Croom Helm, Beckenham

Cosgrove, Dennis 1985 Prospect, perspective and the evolution of the landscape idea. *Transactions of the Institute of British Geographers* 10 45–62

Crang, Mike 2002 Qualitative methods: the new orthodoxy? *Progress in Human Geography* 26 647–655

Crang, Mike 2003 Qualitative methods: touchy, feely, look-see? *Progress in Human Geography* 27 494–504

Cranny-Francis, Anne 2009 Touching film: the embodied practice and politics of film viewing and filmmaking. *Senses and Society* 4 163–178

Crouch, David & Toogood, Mark 1999 Everyday abstraction: Geographical knowledge in the art of Peter Lanyon. *Ecumene* 6 72–89

Daniels, Stephen 1984 Human geography and the art of David Cox. *Landscape Research* 9 14–19

Daniels, Stephen & Cosgrove, Dennis E 1988 Introduction: iconography and landscape. *in:* Cosgrove, D E & Daniels, S (eds.) *The iconography of landscape: essays on the symbolic representation, design and use of past environments.* Cambridge University Press, Cambridge 1–11

Deutsche, Rosalyn 1988 Uneven development – public art in New York City. *October* 3–52

Doel, Marcus A 1999 *Poststructuralist geographies: the diabolical art of spatial science.* Edinburgh University Press, Edinburgh

Doel, Marcus 2000 Un-glunking geography: spatial science after Dr Seuss and Gilles Deleuze. *in:* Crang, M & Thrift, N (eds.) *Thinking space.* Routledge, London 117–135

Driver, Felix 1995 Visualizing geography: a journey to the heart of the discipline. *Progress in Human Geography* 19 123–134

Driver, Felix 2003 On geography as a visual discipline. *Antipode* 35 227–231

Edensor, Tim 2000 Walking in the British countryside: reflexivity, embodied practices and ways to escape. *Body and Society* 6 81–106

Featherstone, Mike 2010 Body, image and affect in consumer culture. *Body and Society* 16 193–221

Foster, Kate & Lorimer, Hayden 2007 Some reflections on art-geography as collaboration. *Cultural Geographies* 14 425–432

Garrett, Bradley L 2010 Videographic geographies: Using digital video for geographic research. *Progress in Human Geography* 35 521–541

Gibbs, Anna 2009 After affect: sympathy, synchrony and mimetic communication. *in:* Gregg, M & Seigworth, GJ (eds.) *The affect theory reader.* Duke University Press, Durham, NC; London 186–205

Giorgi, Amedeo 1992 Description versus interpretation – competing alternative strategies for qualitative research. *Journal of Phenomenological Psychology* 23 119–135

Giorgi, Amedeo 2009 *The descriptive phenomenological method in psychology: a modified Husserlian approach* Duquesne University Press, Pittsburgh

Hawkins, Harriet 2011 Dialogues and doings: Sketching the relationships between geography and art. *Geography Compass* 5 464–478

Hawkins, Harriet 2013 Geography and art. An expanding field: Site, the body and practice. *Progress in Human Geography* 37 52–71

Hawkins, Harriet 2015 Creative geographic methods: knowing, representing, intervening. On composing place and page. *Cultural Geographies* 22 247–268

Hogan, Susan 2009 The art therapy continuum: a useful tool for envisaging the diversity of practice in British art therapy. *International Journal of Art Therapy* 14 29–37

Hogan, Susan & Pink, Sarah 2012 Visualising interior worlds: interdisciplinary routes to knowing. *in:* Pink, S (ed.) *Advances in visual methodology.* SAGE, London 230–248

Jacobs, Jessica 2013 Listen with your eyes; towards a filmic geography. *Geography Compass* 7 714–728

Jacobs, Jessica 2015 Visualising the visceral: Using film to research the ineffable. *Area.* doi: 10.1111/area.12198

Knoblauch, Hubert 2012 Videography. *in:* Knoblauch, H, Schnetter, B & Raab, J (eds.) *Video analysis: methodology and methods: qualitative audiovisual data analysis in sociology,* 3rd edition. Lang, Frankfurt; Oxford 69–83

Lafrenière, Darquise & Cox, Susan M 2013 'If you can call it a poem': Toward a framework for the assessment of arts-based works. *Qualitative Research* 13 318–336

Laurier, Eric & Philo, Chris 2012 Natural problems of naturalistic video data. *in:* Knoblauch, H, Schnetter, B & Raab, J (eds.) *Video analysis: methodology and methods: qualitative audiovisual data analysis in sociology,* 3rd edition. Lang, Frankfurt; Oxford 181–190

Lorimer, Jamie 2010 Moving image methodologies for more-than-human geographies. *Cultural Geographies* 17 237–258

Mackenzie, A. Fiona D & Taylor, Sue Jane 2006 Claims to place: The public art of Sue Jane Taylor. *Gender, Place and Culture* 13 605–627

Madge, Clare 2014 On the creative (re)turn to geography: poetry, politics and passion. *Area* 46 178–185

Marston, Sallie A & De Leeuw, Sarah 2013 Creativity and geography: toward a politicized intervention. *Geographical Review* 103 III–XXVI

Massey, Doreen B 2005 *For space*, SAGE, London

Matless, David & Revill, George 1995 A solo ecology: the erratic art of Andy Goldsworthy. *Ecumene* 2 423–445

Merchant, Stephanie 2011 The body and the senses: Visual methods, videography and the submarine sensorium. *Body and Society* 17 53–72

Mondada, Lorenza 2012 Video recording as the reflexive preservation and configuration of phenomenal features for anlysis. *in:* Knoblauch, H, Schnetter, B & Raab, J (eds.) *Video analysis: methodology and methods: qualitative audiovisual data analysis in sociology*, 3rd edition. Lang, Fankfurt, Oxford 51–67

Morris, Nina J 2011 Night walking: Darkness and sensory perception in a night-time landscape installation. *Cultural Geographies* 18 315–342

Morris, Nina J & Cant, Sarah G 2006 Engaging with place: Artist, site-specificity and the Hebden Bridge Sculpture Trail. *Social and Cultural Geography* 7 863–888

Nash, Catherine 2000 Performativity in practice: Some recent work in cultural geography. *Progress in Human Geography* 24 653–664

O'Neill, Maggie 2008 Transnational refugees: The transformative role of art? *Forum Qualitative Sozialforschung* 9 article 59

Oliver, Stuart 2003 Geography's difficult engagement with the psychological therapies. *Social & Cultural Geography* 4 313–320

Osborne, Brian 1988 The iconography of nationhood in Canadian art. *in:* Cosgrove, DE & Daniels, S (eds.) *The iconography of landscape: essays on the symbolic representation, design and use of past environments*. Cambridge University Press, Cambridge 162–178

Parr, Hester 2006 Mental health, the arts and belongings. *Transactions of the Institute of British Geographers* 31 150–166

Parr, Hester 2007 Collaborative film-making as process, method and text in mental health research. *Cultural Geographies* 14 114–138

Pauwels, Luc 2012 Contemplating the state of visual research: an assessment of obstacles and opportunities. *in:* Pink, S (ed.) *Advances in visual methodology*. SAGE, London 248–265

Philo, Chris & Parr, Hester 2003 Introducing psychoanalytic geographies. *Social & Cultural Geography* 4 283–293

Pile, Steve 2010 Emotions and affect in recent human geography. *Transactions of the Institute of British Geographers* 35 5–20

Pink, Sarah 2009 *Doing sensory ethnography* SAGE, London

Pink, Sarah 2011a Images, senses and applications: Engaging visual anthropology. *Visual Anthropology* 24 437–454

Pink, Sarah 2011b A multisensory approach to visual methods. *in:* Margolis, E & Pauwels, L (eds.) *The SAGE handbook visual research methods.* SAGE, London 601–615

Pink, Sarah 2011c Sensory digital photography: Re-thinking 'moving' and the image. *Visual Studies* 26 4–13

Pink, Sarah 2012 Advances in visual methodology: an introduction. *in:* Pink, S (ed.) *Advances in visual methodology.* SAGE, London 3–17

Pink, Sarah, Hogan, Susan & Bird, Jamie 2011 Intersections and inroads: Art therapy's contribution to visual methods. *International Journal of Art Therapy: Inscape* 16 14–19

Pink, Sarah; Leder Mackley Kerstin, & Moroşanu, Roxana 2015 Researching in atmospheres: Video and the 'feel' of the mundane. *Visual Communication* 14 351–369

Prince, Hugh 1988 Art and agrarian change, 1710–1815. *in:* Cosgrove, DE & Daniels, S (eds.) *The iconography of landscape: essays on the symbolic representation, design and use of past environments.* Cambridge University Press, Cambridge 98–118

Rose, Gillian 2003 On the need to ask how, exactly, is geography "visual"? *Antipode* 35 212–221

Thornes, John E 2004 The visual turn and geography (Response to Rose 2003 intervention). *Antipode* 36 787–794

Thrift, Nigel J 1996 *Spatial formations* SAGE, London

Thrift, Nigel J 1997 The still point. *in:* Pile, S & Keith, M (eds.) *Geographies of resistance.* Routledge, London 124–151

Tolia-Kelly, Divya P 2007 Fear in paradise: the affective registers of the English Lake District landscape revisited. *Senses and Society* 2 329–352

Tolia-Kelly, Divya P 2012 The geographies of cultural geography II: Visual culture. *Progress in Human Geography* 36 135–142

Wylie, John 2005 A single day's walking: Narrating self and landscape on the South West Coast Path. *Transactions of the Institute of British Geographers* 30 234–247

Wylie, John 2006 Depths and folds: on landscape and the gazing subject. *Environment and Planning D: Society & Space* 24 519–535

Wylie, John 2011 Landscape. *in:* Agnew, J & Livingstone, D N (eds.) *The SAGE handbook of geographical knowledge.* SAGE, Los Angeles; London 300–315

Zukin, Sharon 1989 *Loft living: culture and capital in urban change* Rutgers University Press, New Brunswick, NJ

Zukin, Sharon 1995 *The cultures of cities* Blackwell, Oxford

PART II

Exploring Gendlin's Ideas through Artistic Practice

CHAPTER 3

Implying and Occurring

Abstract In this chapter, Banfield provides a focused account of Eugene Gendlin's philosophy, and addresses a core philosophical concern within non-representational geography: human subjectivity. Janet explains Gendlin's core concepts and complex terminology (e.g. implying, occurring, interaffecting) and illustrates them in the context of empirical data from geographical research into artistic spatial experiences. She addresses specific themes of geographical interest—space, time and subjectivity—and makes a valuable intervention to contemporary debates within non-representational geography concerning the value of, need and capacity for a degree of humanism despite the anti-essentialist nature of non-representational thinking, by outlining how Eugene Gendlin's philosophy accommodates a sense of a personal past, the future and personal agency.

Introduction

Gendlin's philosophy has relevance for geography, and I illustrate and explore this philosophy in the context of my own research. Except where otherwise specified, all references to Gendlin's work in this section refer to his text *A Process Model* (Gendlin 2001), and I have organized this discussion around three broad themes: interaffecting and coordinated differentiation; bodily implying and "had" space-time; and temporality and agency. Gendlin develops a particular and complex vocabulary, which I introduce in a staged approach, outlining Gendlin's ideas in his own words in the

© The Author(s) 2016
J. Banfield, *Geography Meets Gendlin*,
DOI 10.1057/978-1-137-60440-8_3

first instance, before drawing out empirical illustrations of these ideas from my research to make this terminology more accessible.

My starting point for this empirical illustration of Gendlin's philosophy is an account provided by Jane of an occasion when she was living and working in China, during which she painted a traditional Chinese homestead. Jane described a ten-year-old girl approaching her as she sketched the handle of an axe that was lying on the ground. As the original line she drew was incorrect, Jane then drew a new line, at which point the girl "picked up my eraser and very gently she rubbed out the incorrect line of the axe". Shortly afterwards, when Jane's pen lid dropped to the ground, the girl picked it up, blew the dust off it and "gently put that back on my felt pen". The girl's mother, whom Jane had recently drawn in the distance, then came over to see what was happening, at which point the girl indicated in gestures that Jane had not included her mother's curved cutting knife (which was attached to her belt) in her painting. Jane said that she had not really noticed the belt at a distance but that the girl's actions indicated the importance of the knife to their agricultural community, and that she immediately painted the knife into her picture. She described being touched at the girl's gentle concern and said that "it told me something about their community because she was accepting me".

Interaffecting and Coordinated Differentiation

As outlined in Chap. 1, two core ideas are central to Gendlin's ethological model, implying and occurring, which are distinct but interactive. Implying is ongoing, and more is implied than occurs. Occurring is change; something happening. What occurs always occurs into implying because implying is ongoing, and occurring can change that implying. In essence, the implicit can be thought of as the whole range of interconnected events and things that could occur; implying is a kind of stimulus for change; and occurring is a change that happens. The process/interaction is both the implying and the occurring. Any occurring implies further occurring, and can change what is implied. As implying implies (stimulates) but does not determine what occurs, implying also stimulates its own change, generating a future that is not predetermined. Gendlin's term for this multiple cross-influence between implyings and occurrings is "interaffecting" (Banfield 2014).

We can both illustrate these terms and pursue Gendlin's thinking further by relating these ideas to Jane's account of painting the Chinese homestead. In Jane's account the artistic process/interaction implied the painting of the scene and its experience, but not in any predetermined manner. The implying implies (stimulates) some way of carrying forwards, but not any one particular way. In this case, the axe on the ground implied its own drawing, but the occurring of its being drawn was not as originally implied, requiring (and implying) remedial action by Jane. In drawing a new line, the implying occurred, while the girl's actions in picking up the eraser and rubbing out the incorrect line occurred into the ongoing implying of Jane's practice. The implying and occurring of this interaction further implied the occurring of the mother's entrance to the interaction, illustrating Gendlin's interaffecting, which drew progressively more people into the artistic practice process.

In keeping with other non-representational philosophies, Gendlin depicts a transversal interconnectedness that precedes ontological distinctions. Gendlin describes this as an original interaffecting, or eveving (the interaffecting of *ev*erything by *ev*erything). Gendlin speaks of an "implicit many", from which processes and entities are separated out. This emphasizes the multiplicity of possible implyings and occurrings and their implicit connectivity (their interaffecting). This interaffecting of the implicit precedes the implicit being many. In other words, things become many as they separate out. Gendlin's term for this separating out of processes and entities is coordinated differentiation. By this, Gendlin means that processes and entities separate only in a coordinated way, with each developing only together with some others and only without some others. The exact way that one process or entity is, implies how the others are, and each is just so only if the others are also just so. The differentiation of processes and entities occurs in a coordinated manner (Banfield 2014).

The occurring of Jane's artistic practice can be thought of as occurring into (and from within) an original interaffecting or implicit many. As an implicit many, not only are Jane, the girl, the mother, the homestead, the knife, the pen, the eraser and so on already interaffecting one another (mutually connected or influencing) but so too are Jane's family, relations and background, the girl's wider family, community and locality, and so on, in a transversal connectivity within which multiple implyings and occurrings interaffect one another. It was the process of Jane's artistic practice of painting the homestead as a whole that was implied, but different sub-processes only occurred in the manner they did because some other sub-processes occurred as they did. The sub-process in this instance is the drawing of the mother's knife. This only occurred in the manner that it did because the drawing of the axe (through the interaction between Jane and the girl) implied the inquisitive entrance of the mother to the interaction (her occurring into the ongoing implying). If the mother had not approached Jane and the girl, Jane would not have been able to see the knife clearly, nor would the girl have been prompted to indicate that the knife was missing. The absence of the knife was already implied, but the occurring of the girl in pointing out its absence implied its inclusion, an implying which then occurred.

Another aspect of Gendlin's philosophy that comes through in Jane's account is that objects are also a function of this process because the process separates out the objects by not occurring without them (which Gendlin calls a "stopped" process) and resuming if they recur. Once the process resumes, it is no longer implied but occurs. For Gendlin, an object is held stable by the entity's behaviour, such as a mouse remaining constant for the cat that is chasing it. The object is relative to the body-implied behaviour sequence, and is recognized through the resumption of a process with the body (it is brought back into the process). By recognizing objects through their inclusion in a process, it is those objects associated with (and emergent within) the ongoing process that are granted agency. However, this circulation of agency among emerging entities in an ongoing process does not deny the influence or agency of the stopped process, because what does continue does so differently as a result of the stoppage. Gendlin therefore provides for a broad notion of agency by attributing agency to objects recognized by their inclusion in an ongoing process, and to stopped processes without which the ongoing process could not go on in the manner that it does (Banfield 2014).

In Jane's painting of the Chinese homestead, we can think about the coordinated differentiation of objects in relation to the curved cutting knife. As a significant aspect of life on the homestead, the inclusion of the knife in the painting was originally implied by the artistic practice process, even though its inclusion could not occur until Jane's interaction with the girl had drawn the attention and implied the entrance of the mother to that interaction. The object was thereby recognized through its inclusion in the artistic practice process. We can also think about both the inaccurate line in the drawing of the axe and the dropping of the pen lid in terms of Gendlin's stopped, carried and continued processes. The artistic subprocess of drawing the axe as originally implied was temporarily stalled by the drawing of an inaccurate line, while the dropping of the pen lid also stopped the artistic sub-process for a short time. However, the erasure of the inaccurate line implied the drawing of an accurate line, which enabled the resumption or continuance of the sub-process, and the same can be said for the retrieval of the pen lid. In both cases, the stopped sub-process was carried forwards by the ongoing implying of the artistic practice process as a whole, and had agency within that process.

Coordinated differentiation applies not only to processes and objects but also to subjects, in this case Jane, the girl and the mother. These subjects were coordinately differentiated through Jane's artistic practice, which I conceptualize in interrogative terms as Jane repeatedly tested her felt sense of the situation as her artistic practice process progressed. As Jane delved deeper into her subject matter, the subject matter also posed questions of Jane. The occurring of the dropping of the pen lid functioned as an interrogative fulcrum in implying a response from the girl, and the occurring of the response implied a response from the mother. This in turn implied an interrogation on the part of Jane as to her relation to the people and community around her, changing her felt sense of the situation in which she was working and through which she was emerging. This interrogation progressively became a self-interrogation, acting back upon Jane, changing the way in which she saw, experienced and understood both the homestead and its occupants, and herself. Jane's interrogation of the rural community also caused her to interrogate her own understanding of the cultural differences between the painter and the painted, and to evidence that understanding by painting in the knife.

Jane's occurring, in the form of the initiation of her practice, implied the whole of that practice process, through which Jane, her materials, the girl, the mother and the knife became coordinately differentiated. This emphasizes the interaffecting of everything by everything and the coordinated differentiation of subjectivities and spatialities as artistic practice unfolds, and brings us to a point at which we can begin to think about spatiality in Gendlin's work.

BODILY IMPLYING AND "HAD" SPACE-TIME

In relation to the body's own implying, Gendlin describes this as focaling, which echoes Massumi's (2002) notion of sensation as channelling a field of potential into a local action, except (as I flesh out in the next section) that Gendlin's focaling seemingly allows for greater personal influence on what is implied or channelled. This focaling in effect pulls together the stimulus for change from multiple implyings into one implying, enabling the body to enact its own behaviour sequence. As such, the bodily implying includes a whole context of mutually implicit behaviour sequences that are focaled with the actual environment that happens into it. From this, Gendlin develops his notion of behaviour space as a mesh of possible behaviours that the body implies in all directions and respects. Here, Gendlin's work is not dissimilar to that of Bergson's (1911) zone of indetermination surrounding a living being in its activity and giving an estimate of those things with which it is in relation. Describing this behaviour space as a mesh in which any occurring sequence changes how others would occur if they were to form after it, Gendlin also specifies the concept of "had" space-time, whereby the body feels and perceives the whole context of implicit behaviour possibilities, taking account of other objects in the space as they are recognized in the context of implicit behaviour sequences. As behaviour always occurs in the midst of other implicit behaviours and as a change in those behaviours, the whole mesh (behaviour space) is carried forwards or continued (Banfield 2014).

Thinking back to Jane's experience and account of painting the Chinese homestead, Gendlin's notion of "had" space-time seems quite apt, as Jane's behaviour sequences implied changes in the behaviour sequences of others around her, which implied the recognition of objects (the axe, the pen lid, the curved cutting knife) in the context of those behaviour sequences, and which implied the recognition and constitution of both objects and subjects within the behaviour space. The whole was carried forwards as it was transformed by this interrogative unfolding through the interaffectivity of implyings and occurrings, incorporating more of the community and culture around her as she increasingly became implicitly part of it through her acceptance by the girl.

Laura's earlier description of her experience of being simultaneously in her studio and in her painting also lends itself to conceptualization as "had" space-time, although in a different manner to that of Jane's homestead painting. Laura talks about the place in her painting coming alive under her brush, and a sense of great joy in that experience. Laura's experience of starting to be in her painting through the materials with which she is painting is a highly affective experience, far more than the common visual associations that we tend to make with ideas of paintings, photographs and memories. Working initially from photographs but progressively from her own memories, we can think of these photographs and memories as being occurrings, which change Laura's artistic practice as it was previously implied, implying brush strokes which, in their own occurring, change the ongoing implying further.

Through her practice, Laura experiences both the present and the distant simultaneously, undermining any notion that they are separate realms, consistent with the non-representational unsettling of distinctions between dualisms such as real and ideal. It is in the material practice of her art that the worlds of Laura's studio and her art are one, and it is during her material practice that this is most clearly articulated. This is not a case where distinct or pre-existing realms are brought together, but they come into being as one (Banfield 2014). Laura's practice here is an example of Gendlin's whole process that is interaffectingly one but becomes coordinatedly differentiated. Laura, her photographs, her memories, the paper, the place unfolding before her, her brushes and paints, are not separate entities that are pieced together in a jigsaw-like fashion, but in their interaffecting they each imply the whole and the whole implies each, with their occurrings changing the implying that is ongoing. Laura's practice is an interactional process in which the body-environment concretion changes its environment and goes on in that environment, which Laura experienced as the place in the painting coming alive under her brush. Laura is implicitly aware of possible behaviour sequences appropriate to both the place of her studio and the place that she is painting, but she is implicitly aware of them as a unified whole: it is a unique, situated and implicit "had" space-time.

In Susan's accounts of her artistic practice, too, we can discern "had" space-times, but in yet another way. Susan works in a number of artistic media, some of which exhibit locational specificity, and her interview account addressed each of them. For example, printmaking can only be conducted in her studio in the UK as that is where her press is located, while jewellery-making could be done in either the UK or at her studio in Greece, but is undertaken in Greece because of the beauty of the pebbles on the country's beaches. The material occurrings of her artistic practice, which are specific to the locations of her practice, influence the manner in which Susan responds to those locations, whether the implying of the artistic practice process is itself location specific (print-making) or not (jewellery-making). Spatiality and subjectivity are mediated through Susan's artistic materials and equipment, as articulated particularly clearly in relation to her paintings of Greece.

Susan says that with her paintings of Greece, she does not want to copy the "obvious beauty of the place" but wants to acknowledge the wires, half finished buildings and "rather tatty piers going down to a boat that's a bit shabby". She says that she does not so much see that side of Greece in the landscape as she remembers it in the landscape because "in the bright of sunshine you don't see that, you only see the lovely blue sea and the sky". For Susan, both spatial and temporal distance are beneficial for her paintings of Greece, enabling her to overcome the captivating power of its obvious beauty. Susan indicates a need to distance herself from the tourist ideal of Greece in order to access what she perceives as the reality of Greece. Whereas Susan's appreciation of the beauty of the pebbles in Greece stimulates her jewellery practice in situ, her appreciation of the country's shabbiness stimulates her painting practice only at a distance from Greece. We might associate this with Gendlinian ideas of implying and occurring, whereby the captivating power of the ideal Greece, a powerful affective force, is excessively implied, and to overcome this excess, spatial and temporal distance are sought so that the real Greece can occur. Such relative distancing can perhaps be considered a specific form of coordinated differentiation. The relation between Susan's affective proximity to Greece and her physical proximity to Greece is a dynamic one, generating particular and unique "had" space-times through the coordinated differentiation of her artistic practice process (Banfield 2014).

The presentation of such composite or hybrid experiences in pictorial form can be thought of as opening on to virtual space and time and fabricating a world in itself (Crowther 1993; Radley 1996; Mooney 2002), in other words, Gendlin's "had" space-time. Laura's and Susan's paintings, for example, mapped out the contours of an experiential space which resonates beyond the image content of the painting, incorporating extraneous, manipulated and affective elements (Manning 2009). These artworks are cartographic, in the sense of laying out a spatio-temporal consciousness, but pre-cartographic in that they are concerned less with the coordinates of its form than the experience of its deformation (Carter 2009; Manning 2009). Multiple spatialities and temporalities are experienced together in a complex and contingent interaffectivity through artistic practice. For Jane and Laura one artwork or artistic practice connects them experientially to different space-times, while for Susan different artworks or artistic practices connect her experientially to the same space-time. However, thinking in terms of relationscapes (Carter 2009; Manning 2009) like this sustains the idea of distinctions between different spaces and times, which are subsequently drawn together through artistic practice. In this understanding, Jane's personal background and culture are brought into increasing identification with the cultural community of the girl and the mother through such a relationscape; Laura's landscape painting also serves as a relationscape between different frames of reference; while for Susan it is Greece that could be considered the relationscape, such that its beauty and shabbiness are perceived differently in different artistic practices at different distal removes from Greece. However, this suggests separateness rather than interaffecting (Banfield 2014).

Whereas, in what Paul Carter calls an angelic topology, space-time is folded and deformed, such that the present and absent, the real and ideal, the distal and proximal, the past and future are brought together in a timeless relationscape (Carter 2009; Manning 2009), in Gendlin's "had" space-time all these qualities are already unified through their original interaffecting. Conceptualizing such artistic spatialities and subjectivities as relationscapes suggests that different axes of identification pre-exist the artistic practice through which they become tangible, but in Jane's experience of painting the Chinese homestead, this was not at all the case, as her relations to those with whom she was interacting only emerged through her artistic practice. The individual subjects who emerged became coordinatedly differentiated even as their interaffecting became more acutely felt. Rather than being brought together, qualities such as presence, proximity, reality, relation, are coordinately differentiated through the carrying forwards of the whole process, allowing their recognition as they become progressively differentiated from the implicit many and taken up by the process.

Developing Gendlin's thinking in light of these practices, we can perhaps think productively about the differences in "had" space-times as different forms or registers of coordinated differentiation. Jane, the girl, the mother and the knife became coordinatedly differentiated through the implyings and occurrings that led to the recognition of the significance of the curved cutting knife. For Susan, the real Greece became coordinatedly differentiated from the excessive ideal Greece through the occurring of distance from Greece during her artistic practice. For Laura, it is less clear what precisely is being coordinatedly differentiated, but it seems to combine elements of both Jane's and Susan's practices. Looking at the rest of Laura's account of this work, Laura talks elsewhere about usually not including figures or animals in her work, and says that this work is particularly challenging for her as she needs to include a bird to satisfy the person who commissioned the work. This suggests similarity to Jane's account of gaining an understanding of the homestead community through the implyings and occurrings of her artistic practice. As a parallel, Laura's artistic implyings and occurrings are interrogating the belonging and significance of the bird within its artistic environment, through which she increasingly feels part of that environment. At the same time, and similar to Susan's account, Laura talks elsewhere about working less from the photographs and more from her own experience as the work progresses, distancing herself from the affective power of the photographs' visuality in much the same way that Susan distances herself from the affective power of Greece's visuality. Consequently, Gendlin's ideas of coordinated differentiation in the generation of implicitly "had" space-times accommodate the variety of artist experiences and accounts considered here, in a way and to a degree that sits uneasily with other contemporary concepts through which geographers seek to understand such artistic spatialities and subjectivities.

Exploring these artistic practices (and the spatialities and subjectivities that emerged through them) in the context of Gendlin's philosophical works has both enabled the elucidation and illustration of some of Gendlin's core concepts, and introduced new ideas and terminology to non-representational geography and to the geographies of art. These Gendlin-informed ideas and terms can help geographers to think about and work with artistic practices, experiences, and "had" space-times, which are individual in the sense of being unique, but not personal in the sense of being the property of an essential identity. In generating emergent subjectivities, these artistic practices bring us to a discussion of what Gendlin has to say about temporality, subjectivity and agency, with the potential to make a very specific contribution to non-representational geography.

TEMPORALITY AND AGENCY

Gendlin specifies that the process includes both implying and occurring, and that time is generated by this process, by the relation between them. For Gendlin, something is past, present or future depending on its function in the occurring into implying cycle. The old sequence is implicit in the occurring of the new, and the new happens into the implying of the old. The present process goes on in the body-environment it has generated, such that the body-environment is a past in the present and implying is a future in the present. For Gendlin, how the body-environment functions is the past, and how implying functions is the future; occurring regenerates the body but also changes the implying. Consequently, parallels can be drawn between Gendlin's temporality of the body and Bergson's (1911) idea of the body as an ever-advancing boundary between past and future, where the past expires in a deed. Where for Bergson the past is exhausted and will only recover an influence by borrowing the vitality of present perception, for Gendlin this recovery of influence is accomplished by the crossing of everything that happens with everything that has happened (Gendlin 2009b). The past never fully expires by virtue of its ongoing implyings (Banfield 2014), the continued interaffecting after the cessation of an occurring.

In essence, the present is experienced with, through and by means of the past; the present happens in the remains of the past, and the past functions as already changed by the present it functions in. The present is only as it is because the past is as it was, but the past is only as it is now because the present is as it is now. In other words, the meaning and influence of the past is dependent upon the present. Processes function as already interaffected; through occurring they re-determine their multiplicity, changing the system of possibilities for the cycle. Gendlin's notion of interaffecting is reminiscent of Bergson's (1911) propagation of modifications throughout the immensity of the universe and Massumi's continuity of transitions (Massumi 2002), in that the interaction determines how each component acts. Interaction is a single undivided system. In Massumi's virtual swarm of potential configurations there subsists an infinity of qualitative relational differences or productive interferences, such that it constitutes a continuity of transitions rather than a collection of discrete elements (Massumi 2002). What each is, is already affected by the other. As Gendlin specifies, "what *occurs* is the result of how the effect of each process, part, or difference, is changed by how its effect on the others affects it" (Gendlin 2001: 40, emphasis in original). In a similar manner to the original interaffecting of the implicit many, past, present and future are also interaffected. This enables us to understand Gendlin's specification of two types of implying (horizontal and temporal). Horizontal implying refers to the original interaffecting, through which parts and wholes imply each other. Temporal implying refers to the interaffecting of past, present and future, which are interlocked but not linear.

Nonlinear temporalities can also be discerned in the practices and accounts of a number of participating artists. The clearest example comes from Jane's practice-based account, which articulates both the immediate responsiveness of a present occurring when Jane paints into the scene a shed door that had blown open, but also conveys her temporal shifting during her practice. While responding to the events occurring around her, Jane also brings extensive past experience into her practice by ensuring that certain trees, of which she has fond memories, are afforded prominence in some way. Similarly with respect to the past, Jane describes painting her husband into the scene at a later date in recognition of his work in tending the garden, a temporally extended practice which constitutes the present environment to which she is immediately responding and which is also brought into the present through the functioning of her body in the present. With respect to the future, Jane also evidences implying in its function as the future by leaving a work unfinished for several weeks to allow the fruit on the tree in the orchard to ripen so that it can be depicted in the finished work. However, this also allows the return of the past into the present, as it is in drawing on past experience of the fruiting of the tree that this future fruiting can be implied in the painting. Neither Jane's husband nor the fruit were present in the research session, but in the "had" space-time of Jane's artistic practice, the past was bodily present and the future implyingly present. These implyings and occurrings contribute to granting such artworks a sense of personal identity or subjectivity. It is Jane's artistic practice process and Jane's memories which act as stimuli for the interrogative process through which the implicit many is coordinatedly differentiated, and it is her garden which accommodates trees that she and her husband have known together. This is not to deny the agency of the objects, materials and other entities involved in this artistic interaffecting, but it is to suggest that one of the things that is coordinatedly differentiated through Jane's artistic practice is her sense of identification with these particular aspects of her body-environment.

This raises the issue of the human subject, or sense of a personal self, which constitutes a specific form of a more general concern with how we can account for stasis amid the dynamism of ongoing flux in non-representational philosophies (Massumi 2002; Anderson and Harrison 2010a). How might non-representational theories accommodate a minimal humanism (Thrift 2008; Wylie 2010)? While non-representational theories neither eliminate nor reassert the subject, they employ distributed, performative and relational notions of subjectivity (Wylie 2010), as seen in Jane's paintings. However, if in their emergence subjects are radically contingent and irreducibly specific (Anderson and Harrison 2010a), how can Gendlin help us to account for the phenomenal sense of an enduring identity and a personal future? (Banfield 2014).

In criticizing non-representational thinking for not allowing a vision of a future and its lack of an agentive subject, Rose (2010) compared temporality within the philosophies of Deleuze and Levinas. In the former, envisaging a future is deemed by Rose to be untenable because, although the permanent presence of the past enables the framing of expectation based on past events, the future is fully open so expectation becomes redundant. In the latter, the past cannot be achieved and the future cannot be anticipated, rendering the envisaging of a future possible but futile because it can never be attained. Gendlin's *Process Model* strikes a more optimistic note in terms of the potential for forward thinking and agency. To see how this is so, it helps to consider together what Gendlin has to say in relation to interaffecting and time (Banfield 2014).

As with Deleuze and Guattari, who spoke of a permanent reverberative presence of all past events, deriving an echoic sense of personal continuity (Deleuze and Guattari 2004), Gendlin proposes that all events or occurrings continue to interaffect once they have occurred (Gendlin 2009a). At first glance it might appear that a Gendlinian sense of enduring identity and personal past is also echoic in nature. However, this would under-represent the importance of interaffecting to human subjectivity in Gendlin's work, at least in my reading of it. In Gendlin's *Process Model* it seems to me that the most important features for agency and the persistence of subjectivity are that any occurring changes the subsequent implyings and occurrings, and that what occurs has already been affected by the differences it makes in the others which affect it. Thus the occurring of an emergent entity alters its own chances of persisting through its subsequent implyings and occurrings, providing the potential for endurance, in what we might characterize as a consolidation of interaffecting or a reinforcement of focaling. The body, for Gendlin, does not last in remembered time, but in time which its own changing generates; the changing body generates the time in which it remakes itself as the same (Gendlin 2001). That this interaffecting does not tend towards totalization or stasis results from two factors. The first is that the ongoing affectivity of the past is not a concretization of past events, but a perpetual change in the implyings and occurrings. The second is that any one event is not emergent from the implyings and occurrings of just one pre-emergent entity, but of numerous such pre-emergent entities, all of which feed into the perpetual flux of interaffecting (Banfield 2014).

The key challenge, however, for issues of non-representational humanism is how individuals come to generate not only a sense of a personal past but also that of a personal future, if the future is entirely open. While it was noted above that the interaffecting of the present by the occurrings and implyings of the past provides a means of generating a sense of a personal past through the accumulation of bodily-relevant occurrings and implyings, it is also possible for both a sense of a personal future and a degree of agency to be accommodated by Gendlin's *Process Model*.

Whereas for Deleuze and Guattari the future is wholly open and cause and effect are rendered incalculable (Rose 2010), for Gendlin any process, including the body, implies itself by implying the whole process (Gendlin 2001), as in Jane's initiation of her artistic practice process implying her own emergent subjectivity. The process therefore implies its own, not fully determined, range of future occurrings. The question remains, however, as to how the process can "know" of these possibilities in advance, to have a sense of its own future. This question can be answered by considering Gendlin's notion of eveing, or interaffecting, as a process functioning as already interaffected by how its influences on others influence it; as reverberating. While Gendlin's particular term of eveing may not be to everyone's liking, it is not dissimilar to Massumi's (2002) sense of that which emerges changing the conditions of emergence. This interaffecting of everything by everything also includes the effects on the current process brought about by the future affects implied by different implicit behaviour sequences currently facing the process or body. This generates the capacity for individuals to have an implicit awareness of future potentiality, at least to some degree. Consequently, as processes, bodies not only imply their own, not fully determined, range of future occurrings but, since occurrings and implyings mutually imply each other, bodies are also implicitly, or interaffectedly, aware of the not fully determined range of possible changed implyings that might result from their present occurrings. In this sense bodies are aware that their own occurrings or actions have a capacity (albeit uncertain) to alter future implyings and occurrings (Banfield 2014).

If processes function as already interaffected, implicit awareness of the changed implyings subsequent to a past occurring can also alter the present occurring of an emergent entity because the future implyings have already implied the present. This means that the ongoing interaffecting resulting from a past occurring does not just influence the present occurring through the direct functional sequence of implying–occurring–implying, but also through the present implicit awareness of the possible future changed occurrings and implyings contingent upon the past occurring.

Consequently, in focaling the many parts and processes into one implying or local action, bodies can imply certain occurrings that increase or decrease the relative implying of certain other subsequent occurrings and implyings. In other words, they have a degree of agency. Developing Gendlin's notion of interaffecting to this extent indicates a means by which individuals generate awareness of a personal past, the future and a sense of personal agency (Banfield 2014).

However, neither the range of possible future occurrings nor the range of possible changed implyings is fully determined because occurrings can change the implyings in ways that were not originally implied (Gendlin 2001). Again, Jane's paintings provide examples here. In the event, the fruit did ripen and Jane painted them in, but an unseasonably early frost could have occurred into Jane's implying, preventing the fruit from ripening. By contrast, Jane's husband had been expected to join the production session so that Jane could paint him into her work, but he did not do so. This unexpected occurring, however, did not alter the original implying, as Jane painted him in at a later date, consolidating their interaffecting and their emergent subjectivities. Thus, the implicit sense of future potentiality is neither complete nor certain; the future is open and the interaffecting of everything by everything enables the generation of but a limited and hazy sense of a future self and future actions. While the unpredictability arising from the perpetual flux of interaffecting means that such implicit forecasting is not infallible, the continual interaffecting also means that it is not entirely futile.

In summary then, while the uncertainty and multiplicity of human-environment concretions in the *Process Model* prevents deterministic predictions of the future, the enactment of present events can change the relative affectivities in a not fully predetermined manner, with implications for subsequent implyings and occurrings. It is therefore possible within Gendlinian philosophy both to envision a future, and to strive towards it, albeit with the caveat that the attainability of this future is not guaranteed. Integrating Gendlin's thinking into non-representational geographies seemingly can accommodate a minimal humanism, allowing for a sense of a personal past, personal future and personal agency within a non-essentialist, non-deterministic transversal connectedness (Banfield 2014). As the issue of humanism or the human subject is a matter of such contention in non-representational geography, the potential for Gendlin's philosophical writings to contribute to the resolution of these debates warrants serious critical consideration.

CONCLUSION

This chapter has explored the works and practices of participating artists in the context of Gendlin's philosophy and has illustrated some of Gendlin's core ideas with empirical material, making them available and manipulable for geographic inquiry. This paves the way not only for the potential development of a new stream of non-representational thinking in relation to geographies of artistic practice, but also for a non-representational philosophy that can accommodate a human subject, sense of futurity and agency.

The finished artworks here are not one-to-one correspondences or reflections of their subject matter or their creator, but are the culmination of the implicit or affective experience of their production. They interaffect rather than reflect. "Had" space-times are actively and interrogatively forged in artistic practice by interaction between implyings and occurrings. In a practice that is interaffectingly whole, these "had" space-times are implicitly more than occurs in the representational form of the finished artwork. The finished artwork is coordinatedly differentiated through the process of its creation, alongside its creator, the materials they used, the subject matter concerned and so on. The interaffecting of everything by everything allows for the inclusion of myriad entities and events in the artistic practice process. However, any particular entities and events are only included by virtue of the implying–occurring–implying cycle, and are only recognizable by virtue of the coordinated differentiation through which they separate and become taken up by the process. Yet each of these differentiated elements remains interaffectingly connected to each other and to the implicit many from which they emerged, allowing for their continued interaffecting beyond the completion of the artwork. However, the "had" space-time originally generated is changed by the ongoing interaction of implyings and occurrings in this interaffecting. Artistic practices constitute particularized, contingent and fleeting subjectivities and spatialities as the implicit many becomes coordinatedly differentiated (albeit with the potential for ongoing interaffecting), as epitomized in Jane's painting of the girl, her mother and the knife.

REFERENCES

Anderson, Ben & Harrison, Paul 2010a The promise of non-representational theories. *in:* Anderson, B & Harrison, P (eds.) *Taking-place: non-representational theories and geography.* Ashgate, Farnham 1–34

Banfield, Janet 2014 *Towards a non-representational geography of artistic practice.* Unpublished doctoral thesis, University of Oxford, Forthcoming online: https://ora.ox.ac.uk:443/objects/uuid:dd12e1c4-f222-435b-adc0-c1bb68e4f4ac

Bergson, Henri 1911 *Matter and memory* Swan Sonnenschein; Macmillan, London; New York

Carter, Paul 2009 *Dark writing: geography, performance, design* University of Hawai'i Press, Honolulu

Crowther, Paul 1993 *Art and embodiment from aesthetics to self-consciousness* Oxford University Press, Oxford

Deleuze, Gilles & Guattari, Félix 2004 *A thousand plateaus: capitalism and schizophrenia.* Continuum, London

Gendlin, Eugene T 2001 *A Process Model* The Focusing Institute, New York

Gendlin, Eugene 2009a What first and third person processes really are. *Journal of Consciousness Studies* 16 332–362

Gendlin, Eugene T 2009b We can think with the implicit, as well as with fully-formed concepts. *in:* Leidlmair, K (ed.) *After cognitivism: a reassessment of cognitive science and philosophy.* Springer, London, New York 147–161

Manning, Erin 2009 *Relationscapes: movement, art, philosophy* MIT, Cambridge, MA; London

Massumi, Brian 2002 *Parables for the virtual: movement, affect, sensation* Duke University Press, Durham, NC; London

Mooney, Jim 2002 Painting: Poignancy and ethics. *Journal of Visual Arts Practice* 2 57–63

Radley, Alan 1996 Displays and fragments: Embodiment and the configuration of social worlds. *Theory and Psychology* 6 559–576

Rose, Mitch 2010 Envisioning the future: ontology, time and the politics of non-representation. *in:* Anderson, B & Harrison, P (eds.) *Taking-place: non-representational theories and geography.* Ashgate, Farnham 341–361

Thrift, Nigel J 2008 *Non-representational theory: space, politics, affect* Routledge, London; New York

Wylie, John 2010 Non-representational subjects? *in:* Anderson, B & Harrison, P (eds.) *Taking-place: non-representational theories and geography.* Ashgate, Farnham 99–114

CHAPTER 4

Explication and Sharp Concepts

Abstract This chapter addresses a core methodological concern for non-representational geography: whether and how we can access and apprehend pre-reflective experience. Emphasizing Eugene Gendlin's insistence on our capacity to access and articulate from our pre-reflective experience, which runs counter to conventional non-representational geographical understandings, Banfield explores Gendlin's ideas of explication—the development of reflective (conceptual) understanding from pre-reflective understanding—and sharp concepts, which are both rich in pre-reflective understanding and tightly tied into existing conceptual frameworks. This is a significant and timely contribution to geography's identified needs both for methodological innovation in apprehending affect, and for greater disciplinary capacity to work with images conceptually, by considering the explication of both verbal and visual concepts.

Introduction

In this chapter I continue to draw on interview and practice-based data to explore more of Gendlin's philosophy, attending this time to his ideas about our capacity to explicate our implicit understanding, linguistically and artistically, and in combination. This invites discussion of the relation between language and art, or word and image, which in the context of artistic practice I discuss in terms of narrative and symbolism. Although

© The Author(s) 2016
J. Banfield, *Geography Meets Gendlin*,
DOI 10.1057/978-1-137-60440-8_4

different participants articulated very different perspectives and stylistic approaches with regard to narrative in their work, a key feature of narrative in their artistic practices is its intertwining with symbolism. This enables me to explore Gendlin's work on the explication of sharp concepts (Gendlin 2006)—those that are both formally explicit within a conceptual framework and richly implicit—in the context of artistic practice or image-making. This in turn invites us to rethink disciplinary debates concerning the accessibility of affective or pre-reflective aspects of experience, and to explore Gendlin's ideas and methods as a potential means of addressing emerging concerns about geography's capacity to engage conceptually with images (Hawkins 2015).

EXPLICATION

Two core features of Gendlin's philosophical and psychotherapeutic work are a commitment to our capacity to access and apprehend implicit understanding, and an effort to develop means by which we can achieve this. The term that Gendlin uses for the apprehension and articulation of implicit understanding is explication, or the making explicit of that which was formally implicit. Gendlin's philosophical writings and clinical practice address this potential for the explication of implicit understanding in terms of both narrative, or linguistic, and symbolic, or artistic, modes of explication.

Narrative/Linguistic

One implication of Gendlin's work is that we have come to rely so heavily on formal, logical concepts that we have largely become oblivious to the implicit that lies behind them and from which they emerged. In one sense, the implicit has been dispossessed by the emphasis on logic and rationality, and the task now is to repossess the implicit to enrich and develop our conceptual understanding from a stronger implicit base. Gendlin proposes that we have the capacity to connect intentionally with our implicit understanding to generate new formal concepts from our implicit understanding and to enhance our implicit awareness of existing formal concepts. He identifies four key techniques for doing so: focusing, thinking-at-the-edge, dipping and crossing (Gendlin 1993, 1995, 2009b). As previously outlined, focusing is a technique designed to enhance our sensitivity to the implicit, and thinking-at-the-edge aims to connect our implicit understanding to more formal conceptual frameworks. Relatedly, dipping is a way to return to the implicit, repeatedly and for brief periods, in relation to a specific concept or concern to test and refine our implicit understanding of it. Crossing is a means by which the implicit, as well as the logical, relations between formal concepts can be explored and enhanced, by considering multiple concepts conjointly and asking what each means in the context of the other. In crossing, Gendlin proposes that we can generate new implicit understanding from between concepts, from the more-than-logical that connects them.

Through these means, Gendlin proposes that we can develop our implicit understanding of existing formal concepts and develop new formal concepts directly from the implicit. Importantly, working from the explicit to the implicit need not be confined to identifiable words or ideas but can also take as its starting point pauses, gestures and stutters, as there is as much implicit meaning in the linguistic gaps and dead-ends as there is in eloquently articulated phrases (Gendlin 1993, 1995, 2006). Equally, working from the implicit to the explicit need not generate formal concepts or recognizable words immediately, since initial utterances are progressively refined as they are increasingly related to other formal concepts. The important point is that this refinement remains connected with the implicit understanding from which it emerged. This is essential if the implicit meaning is to be carried forwards in and through the explicated term, hence the importance of dipping. As a result, such concepts would be sharp, in the sense that they are full of both explicit and implicit understandings; they are optimally related to other formal concepts in our logical understanding of the world, and they are optimally connected with our implicit understanding (Gendlin 2006).

Symbolic/Artistic

With regard to the apprehension of implicit understanding, art has been described as infusing the artist with experiential understanding, or of bringing together the sensual and conceptual (Tuan 1975; Crowther 1993; O'Sullivan 2001; Deleuze 2005; Hoekstra 2007; Sullivan 2010; Ingold 2011b), suggesting an affinity between artistic practice and Gendlin's methods (Banfield 2014). Gendlin acknowledges such affinity in his own writings, asserting that focusing is often more effective if an image is allowed to form from implicit understanding before words are found to articulate it, and that working with imagery is powerfully enhanced if we employ focusing (Gendlin 1980). Although Gendlin is not referring directly to artistic practices and products, art as a form of image-making lends itself to application with focusing. A more formal relation between focusing and art has also been developed for clinical application in art therapy, which reportedly gains its therapeutic value by providing concrete expression of the implicit or felt sense (Rappaport 1998; Ikemi et al. 2007; Banfield 2014).

As outlined in Chap. 2, art or image-making can be considered a practice that captures the sensual or ineffable through an implicit attunement to the world to render it in artistic form (Bondi 2005; O'Neill 2008; Hogan 2009; Pink et al. 2011; Hogan and Pink 2012). As such the practitioner is described as being drawn into the world along paths of observation, while also drawing it out in gestures of description (Manning 2009; Ingold 2011a), paralleling Gendlin's ideas of explication by connecting with the implicit to articulate from it, in a visual as much as a verbal form. The suggestion here, then, is that artistic practice can facilitate both connecting with the implicit and explicating the implicit, and that artistic or image-based explication might be more effective than linguistic explication. Indeed, qualitative researchers employing artistic practice as a research method have reported such distinctions between verbal and visual means of dealing with intense emotional experiences, for example, when an inability to resolve an emotional struggle in a painting revealed the full force of an emotional conflict that had not been recognized previously through conversation (Hogan 2003, in Hogan and Pink 2012).

Certain aspects of participants' accounts also suggest if not a firm belief in the capacity to express the ineffable more effectively through art than language, then at least an openness to this possibility. Susan, for example, discussed a series of artworks about a period of ill health, saying that they "would probably be more expressive emotionally than probably I am". Susan says that she had previously found it too difficult to talk about that experience, but that these paintings "were so in your face and so up front that actually they did become me in that I could talk about it". Susan's paintings became her voice when she exhibited them. Significantly, Susan talks about her artistic practice facilitating subsequent linguistic communication, consistent with Gendlin's suggestion that words can sometimes be found more readily subsequent to imagery (Gendlin 1980).

The relative capacity for linguistic as opposed to artistic explication can also be considered in relation to a reflexive example of my own experience of the application of image-making as a therapeutic tool during psychological treatment following a traumatic accident. Initially, the treatment was conversationally based, but several sessions into the treatment, I remained unable to articulate verbally anything of my distress, although I could and did display it emotionally with ease. The distress was clear but the underlying issue was elusive. Not only could I not articulate it, I was not even aware of what it was that I was unable to articulate. As my hobby engagement with art since childhood had come up in an early session, I was tasked at the end of another linguistically unproductive session with trying to draw what it was that I could not say.

I was utterly unconvinced that this approach would work. If I did not know what it was that I could not say, how could I possibly know what it was that I needed to draw? Figure 4.1 shows my drawing.

The drawing did not happen all at once, but in two stages. Most of it came easily. What looks like a figure in a box in the centre is the detail of what happened and the circumstances in which I found myself. That was straightforward; I had no reticence or difficulty in depicting being trapped, pinned down and unable to move, and being crushed. The heavy black marking around the upper edge of the circle (fury) and the lighter, more expansive marking around the bottom edge of the circle (despair) were equally easy to put down on paper. I had no difficulty in communicating visually my sense of helplessness in knowing that there was nothing

Fig. 4.1 Trauma drawing (*Source*: Author (2007))

that I could do to save myself, nobody else was in a position to help me, and that if something did not happen to change the situation I would be killed. Then I stopped.

I kept looking at the drawing and thinking, "that's not it, that's not it, there's something missing", but for some time I could not formulate what it was that was missing. I felt disappointed that this effort had not worked, and I was emotionally exhausted. While doing the drawing had not been difficult, dealing with the emotion that accompanied it had been. I do not quite know how it came to be that the thing that needed to be grasped could suddenly be grasped, and to say that I grasped it is misleading, as that would suggest that I had formulated an idea or sense of something capable of being grasped. I can only describe it as a growing compulsion to "just draw", even though I still did not know "what" to draw. I finally picked up my pencil again, without knowing what was going to come of it. What came was one small addition, shown in detail in Fig. 4.2.

Through my numerous post-accident engagements with medical professionals, it has been variously termed a near-death experience, an out-of-body experience, an existential crisis, and a peritraumatic dissociative episode. None of these terms has any resonance with my felt sense of that experience, the details of which I choose not to discuss here. The details associated with Fig. 4.2 are not particularly relevant, as the key point that I want to draw out is the significance of doing the drawing in the emergence

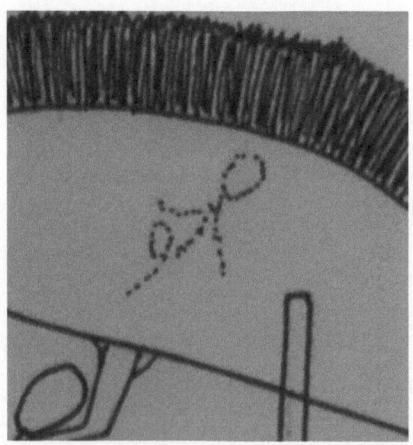

Fig. 4.2 Detail from trauma drawing (*Source*: Author (2007))

of what it was that needed to be drawn. While I still find it difficult to talk (and write) about this event, I now do both, but I had not been able to do so before drawing this image.

Clearly, drawing any substantive conclusions on the basis of so few anecdotal cases, especially if one of them occurred nearly ten years ago, is problematic. I also recognize the limitations of self-reported experiences such as this, although this event took place long before I had encountered the work of Gendlin, or even heard the term non-representational. It is, ultimately, this experience which drew me back to studying, first in psychology and then returning to geography, and to non-representational thinking. However, such examples do suggest sufficient potential for the explication, both artistically and linguistically, of implicit understanding to warrant greater consideration within geographical efforts to apprehend affect.

Drawing together the sensing of implicit sensibilities and qualities, and the seemingly enhanced articulation of the ineffable afforded by its explication through artistic practice, experiences such as those outlined in this chapter suggest that artistic practice might act as a medium for the explication of implicit understanding as both an alternative to, and an intermediary step towards, linguistic communication. In my particular experience, the act of drawing seemed to function as a form of focusing, enabling me to establish a connection with elusive aspects of my implicit understanding in a way that, with time, allowed that implicit understanding to be explicated, artistically in the first instance and subsequently linguistically. Not only might focusing be more effective through artistic practice, but artistic practice or image-making might aid explication of the implicit without deliberate efforts at focusing. This, I think, both reinforces and develops previous accounts of the relations between artistic practice and the implicit (Banfield 2014). On the one hand, Susan's linguistic explication following her artistic explication reinforces Gendlin's assertion that focusing is often more effective if an image is allowed to form from felt sense, only then are words found to articulate it (Gendlin 1980). On the other hand, considering artistic practice as a *form of* focusing develops previous understandings, which saw artistic practice and focusing as related but separate processes (Gendlin 1980), and in which artistic practice occurs subsequent to felt sense (Gendlin 1980; Rappaport 1998; Ikemi et al. 2007). Not only can focusing be more effective through artistic practice, but artistic practice can aid explication of the implicit without deliberate efforts at focusing, whereby artistic practice is its own means of focusing, facilitating the generation of Gendlin's sharp concepts.

Sharp Concepts

Underlying Gendlin's methods for explicating implicit understanding is the idea of sharp concepts. These concepts are both conceptualized formally, for example within an existing structure of language or logic, and firmly rooted in implicit understanding or felt sense (Gendlin 2006). They are rooted in our implicit understanding and meaningfully connected to other formalized concepts. Sharpness provides an interesting way of conceptualizing Manning's form-taking of concepts (Manning 2009), in which sharpness needs to apply to both the implicit and explicit modes of understanding to which the concept relates (Banfield 2014). When a concept is considered to be maximally integrated into a formal knowledge system and also to convey all the implicit meaning with which it is associated and from which it emerged, it is said to be sharp. The concept has to take form in both implicit and explicit understandings.

Initial efforts at explicating from implicit understanding can take the form of noises or nonsense words rather than formal units of language, which might be sharp in the implicit sense but not at all sharp in the explicit sense. My post-traumatic emotional outpourings would fit this description; I was full of felt sense but could not articulate it. By contrast, much of our daily use of language employs our conceptual knowledge in a formal sense and is therefore explicitly sharp, but without any direct sense of its implicit meaning, so is not sharp in the implicit sense. The medical classifications of my trauma experience, which hold no felt sense for me whatsoever, would fit this description. It is in this sense that conventional use of language has largely become isolated from the implicit meaning from which it originally sprang, giving us a false sense of linguistic independence from its implicit roots.

Gendlin's explicatory methods are geared towards linguistic explication. Although, as detailed in the previous section, Gendlin also allows for artistic explication—for example, by advocating linguistic explication following the formation of a visual image—the visual aspect of explication ultimately remains oriented towards linguistic explication. In other words, the image is allowed to form in order to generate a linguistic account of the implicit understanding. This raises at least two issues worth considering in some detail. The first concerns the role of artistic practice relative to artistic product or representational content in the explication of implicit understanding and, related to this, how we might employ artistic practice as a research method in our efforts to access and apprehend the implicit.

I address this first issue in Part 3, where I detail my own experimental application of Gendlin's explicatory techniques in research with practising artists. The second issue is the potential for thinking conceptually about visual images and artworks in Gendlinian terms. In the rest of this chapter, I explore and problematize the idea of sharpness in relation to artistic concepts by considering the variable intelligibility of narrative and symbolism in the practices of participating artists. I then suggest avenues of inquiry through which this might be further investigated, before moving on in the next chapter to consider Gendlin's thinking in relation to the capacity to think from the progressions between concepts, both linguistic and artistic.

Harriet Hawkins has recently commented that, despite geography's characterization as a visual discipline, geographers are perhaps less conceptually skilled with images than with words (Hawkins 2015). This is an interesting observation in light of Gendlin's suggestion that the formation of images prior to linguistic explication aids that subsequent linguistic explication and the generation of sharp concepts. The value of imagery in supporting linguistic explication perhaps suggests a greater intimacy between image-making and implicit understanding than between sentence-making and implicit understanding and, by association, that images and image-making are less conceptual or in some way closer to the implicit than language. However, the capacity to generate linguistic concepts from and through artistic ones indicates considerable conceptual potential within images and image-making for geographical understanding and practice. Collectively, through these theoretical, experimental and critical engagements with Gendlin's philosophy and methods, I hope to make a contribution towards developing image-making as a conceptual form within geography (Hawkins 2015).

The idea of sharp concepts is fairly straightforward in terms of linguistic explication, as formal structures already exist into which emergent implicit understanding can be integrated. There remain, of course, questions as to the translatability of implicitly meaningful understandings between different language systems, and the degree to which implicit sense can be intersubjectively shared to reach a consensus that an emergent concept is sharp in both senses (implicit and explicit). However, if the person generating the implicit understanding is confident that the resulting concept is sharp in both senses, then there is at least potential for another person who understands the resulting concept explicitly to gain an implicit sense of the same through their own felt sense, which is interaffectingly connected with the implicit many.

The issue of conceptual sharpness is perhaps more complicated in relation to artistic explication. In addition to the uncertainty as to intersubjective implicit understanding, different art schools and traditions, both within and across different art media, might function as different language systems in potentially complicating communicability of emergent implicit understanding between individuals and groups. In addition, and perhaps to a greater degree within artistic than linguistic practice, there is the issue of deliberately hampering intelligibility. In linguistic terms, we can think about the challenge of deciphering riddles, while in artistic terms we can consider the accessibility of the meaning of an artwork (its conceptual content) through the artist's varied uses of narrative and symbolism.

While some artists deliberately construct a narrative and present it to the viewer to read in a straightforward fashion, others obstruct narrative in their work, inviting the viewer to generate their own. In her interview, Susan provides an example in which she drew on third-century Middle Eastern fables relating to vice and greed, and used this historic narrative to inform her own narrative about the same themes in current affairs. By contrast, Laura says that she does not inject narrative into her works because viewers bring their own narrative to a painting, reading into it what they will. Clare goes further in her interview, saying that she teases the viewer by deliberately confusing any obvious narrative. Clare says that one body of her work is characterized by humorous other-worldly images of figures doing things that cannot quite be discerned, making people "not quite sure what they're doing, what the story is behind it", and the other body of her work as having only the shape of a moving form but no background, "no content in a way, there's no story". Whether through the unnaturalness of the positions of the figures or the absence of a contextualizing background, Clare deliberately obscures narrative in her work. The symbolism lies in the unnaturalness of the figures and their postures, and it is this symbolism that confuses the reading of narrative in the artwork (Banfield 2014). The intention is not to achieve sharpness but to sustain fuzziness, which raises interesting questions in relation to Gendlin's notion of sharpness, which I explore later in the chapter.

The fuzziness of Clare's narrative has a parallel with Susan's work in which she—uniquely among these participants—talks about intentionally developing a personal set of symbols in her artistic practice, which she says are about her experiences. These symbols are meaningful to Susan but communicate little of her personal experience to viewers. This was evidenced in her interview account of an inquiry from a previous customer

for a second artwork from the same series, which they identified as show-
ing "what looks like a tadpole and a mobile phone". Susan's comment
that the symbols "had nothing to do with a mobile phone and tadpole, for
me those symbols meant something else", illustrates the personal specificity of this symbolism and parallels the impenetrability of Clare's work
due to the lack of intersubjective understanding, narrating a story that
only Susan can comprehend (Banfield 2014). Both narrative and symbolism, then, are not only diversely employed among artists but multiply and
flexibly employed by individual artists, making it especially problematic
to determine sharpness in either an implicit or explicit sense. Susan has
explicated her own implicit understanding into a formalized set of symbols, but as the symbols are unintelligible to other people, nobody else is
in a position to grasp their meaning, either explicitly or implicitly. We can
grasp them, as Susan's customer did, in the context of their closest visual
referent (a tadpole and a mobile phone), yet this gives us no meaningful
explicit understanding and consequently no explicit route into our own
implicit understanding in an effort to establish equivalence of experience.

In addition, artistic materials function in a symbolic fashion and com-
plement the iconographic symbolism in these artists' practices. If we consider together Susan's utilization of Middle Eastern fabulist symbolism
to narrate her own story, and her earlier account of the beauty of the
beach pebbles in Greece, we can conceive the former as iconographic symbolism, and the latter as material symbolism in which the physical entity
of the pebbles also functions iconographically. As the pebbles resemble
the things that they are, the pebbles are concurrently iconographic and
material symbols of the environment of Greece. In the artistic practices
encountered here, symbolism is more-than-visual (Banfield 2014).

Symbolism in artistic practice, then, can take multiple forms (iconographic, material); different approaches can be employed to both symbolism (employing existing symbol systems or developing one's own) and
narrative (constructing or obstructing); and narrative can work through
symbolism (as in Susan's construction of narrative through the employment of fabulist iconography and her obstruction of narrative by inventing
her own symbol system). Interestingly, in Susan's use of Middle Eastern
fables, symbolism works through narrative just as much as narrative oper-
ates through symbolism. By drawing on imagery from historic Middle
Eastern fables that symbolizes vice and greed, Susan narrates her own
perspective on contemporary socio-political issues through historically
and culturally rooted symbolism; her contemporary employment of the

symbolism only makes sense in the narrative context of those historic fables. In these artistic practices, symbolism and narrative are not distinct and isolable, but interpenetrative and mutually constitutive, irrespective of whether that symbolism is iconographic or material in nature and whether narrative is constructed or obstructed (Banfield 2014).

Through the symbolic-narrative evolution of the artwork, the emergent "had" space-time takes material and iconographic form, while the obstruction or construction of narrative through symbolism determines the accessibility of that "had" space-time to anyone other than its creator. Such obstruction of intelligibility—akin to a riddle which demands but disallows deciphering—also obstructs any potential for intersubjective sharing of implicit understanding or pre-reflective experience, and raises a number of questions with regard to attempts at explicating implicit understanding through artistic practice or image-making. I focus here on two sets of questions, concerned with the appropriateness of sharpness in relation to artistic concepts, and its applicability to matters of practice.

One such set of questions concerns whether it is meaningful to think about sharpness in relation to artistic concepts, and how such sharpness might be defined and ascertained. The potential for a defining aspect of an artist's practice to be the obstruction of intersubjective understanding, whether explicit or implicit, has significant implications here. If explicit fuzziness rather than sharpness is the intended outcome, as in Clare's other-worldly figures or Susan's personal symbolism, then can we consider this fuzziness to be sharp in some way? If, in order to be explicitly sharp, a concept (artwork) must be maximally integrated into an existing conceptual system, would an artwork for which maximal integration is zero integration be explicitly sharp despite remaining outside any existing conceptual system? How does an (un)intelligible narrative relate to explicit sharpness in artistic concepts? How do linguistic and artistic modes of explication relate in terms of sharpness? If Susan explained the meaning of her "tadpole" and "mobile phone" in linguistic terms, would that enable us to understand the artwork in which they feature, either explicitly or implicitly? Might the obstruction of intelligibility itself generate meaningful implicit experience in an artistic equivalent to riddles, and how might we work with this potential?

A related set of issues concerns the applicability of ideas of sharpness to explication through artistic practice rather than image form. How can sharpness be brought to bear on the practices rather than the products of art? Can artistic practice itself be sharp, on either an implicit or explicit basis, or is it only the resulting concept, whether linguistic or artistic, that demands sharpness? Does a lack of explicit sharpness in the image concept matter if the process of producing it generates a sharp linguistic concept? What role or value is there in vagueness, indeterminacy, uncertainty and the nebulous? How do or should linguistic and artistic concepts relate in terms of sharpness? To what extent is it helpful to consider image-making as a practice of conceptual sharpening? How is implicit understanding, which is made available for explication through artistic practice, translated into linguistic explication, and how can we experience and understand that process, and its outcomes, intersubjectively?

One example through which we can consider some of these issues comes from my own effort to work with disciplinary concepts on an implicit basis through artistic practice. In essence, I sought to practise the visual as a means of engaging with my embodied response to a particular idea, treating that visual practice as a thinking-space through which new understanding or knowledge might be generated (Crang 2002, 2003; Thornes 2004; Tolia-Kelly 2007; McCormack 2008b, 2012; Hawkins 2015). I worked with a particular phrase in Felix Guattari's *Chaosmosis* (1995), which seemed to resonate with me; it chimed with my felt sense. I used Guattari's idea of a mutant rhythmic impetus temporarily holding together the diverse components of a new existential entity (Guattari 1995) as a pivot around which to dip into my felt sense, trying to move from the words and their conceptual meaning to my felt sense of their meaning, in the context of my own unfolding research and my own emergent subjectivity from within it. Figure 4.3 shows the artistic outcome of this effort.

Several issues arise in relation to this example, which have implications for how we might work with images and image-making on a conceptual basis and which might inform disciplinary efforts to develop new vocabularies and new types of knowledge through working conceptually with and through images (Tolia-Kelly 2007; Pink 2012; Hawkins 2015). The first concerns whether the literary content to which the image refers is intelligible through the image, and whether either the text or the image gain meaning in the context of the other. A second issue concerns individual differences: how differently might other people have responded implicitly and artistically to this same extract from the same text; and how might disciplinary understandings of key terms be enhanced or developed on a collective or collaborative basis through such methods? A third issue concerns the need to develop disciplinary capability to establish a shared

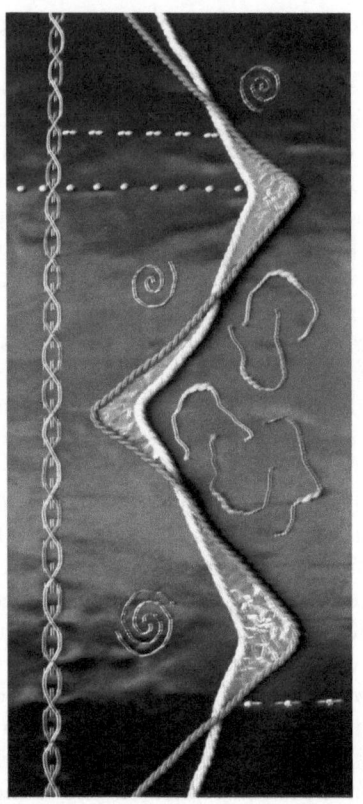

Fig. 4.3   Mutant Rhythmic Impetus Textile and embroidery (*Source*: Author (2013))

understanding on a non-linguistic basis if image-making is to become a conceptual form in geography (Hawkins 2015). A need which might be significantly addressed with further exploration of Gendlin's ideas and techniques, but which also requires resolution of the applicability of sharpness to artistic concepts and practices.

Conclusion

Beginning with a discussion and illustration of Gendlin's ideas on explication in a number of art-related circumstances, this chapter developed into an exploration of my empirical data in the context of Gendlin's notion of sharp concepts. In particular, I drew on the interweaving of narrative and symbolism in artistic practice to complicate how we might operationalize sharpness in relation to images and image-making. This raised a number of issues and questions that start to contribute to efforts to bring practices of image-making to geography as a conceptual form (Hawkins 2015), opening up various avenues for exploration.

The identification of issues surrounding the intersubjective accessibility of implicit and explicit meaning also paves the way for the last conceptual or theoretical element of Gendlin's work that I want to focus on here before moving on to my own experimental application of Gendlin's explicatory methods in Part 3: progressions. In the next chapter I deal with Gendlin's notion of progression between formal concepts, and our ability to "speak" from this between place, which will develop a little further the idea of image-making as a conceptual form.

References

Banfield, Janet 2014 *Towards a non-representational geography of artistic practice.* Unpublished doctoral thesis, University of Oxford, Forthcoming online: https:// ora.ox.ac.uk:443/objects/uuid:dd12e1c4-f222-435b-adc0-c1bb68e4f4ac

Bondi, Liz 2005 Making connections and thinking through emotions: Between geography and psychotherapy. *Transactions of the Institute of British Geographers* 30 433–448

Crang, Mike 2002 Qualitative methods: the new orthodoxy? *Progress in Human Geography* 26 647–655

Crang, Mike 2003 Qualitative methods: touchy, feely, look-see? *Progress in Human Geography* 27 494–504

Crowther, Paul 1993 *Art and embodiment from aesthetics to self-consciousness* Oxford University Press, Oxford

Deleuze, Gilles 2005 *Francis Bacon: the logic of sensation* Continuum, London

Gendlin, Eugene T 1980 Imagery is more powerful with focusing: theory and practice. *in:* Shorr, JE, Sobel, GE, Robin, P & Connella, JA (eds.) *Imagery. Its many dimensions and applications.* Plenum Press, New York; London 65–73

Gendlin, Eugene T 1993 Words can say how they work. *in:* Crease, RP (ed.) Proceedings, Heidegger Conference. Stony Brook, State University of New York

Gendlin, Eugene T 1995 Crossing and dipping: Some terms for approaching the interface between natural understanding and logical formulation. *Minds and Machines* 5 547–560

Gendlin, Eugene T 2006 Transcript of Gendlin Templeton Lecture. Psychology of Trust and Feeling Conference. Stony Brook University. http://www.focusing.org/gendlin_templeton.html. Accessed 05 Nov 2012

Gendlin, Eugene T 2009b We can think with the implicit, as well as with fully-formed concepts. *in:* Leidlmair, K (ed.) *After cognitivism: a reassessment of cognitive science and philosophy.* Springer, London, New York 147–161

Guattari, Félix 1995 *Chaosmosis: an ethico-aesthetic paradigm.* Power Publications, Sydney

Hawkins, Harriet 2015 Creative geographic methods: knowing, representing, intervening. On composing place and page. *Cultural Geographies* 22 247–268

Hoekstra, Daan 2007 The artist's study of nature and its relationship to Goethean science. *Janus Head* 10 329–349

Hogan, Susan 2009 The art therapy continuum: a useful tool for envisaging the diversity of practice in British art therapy. *International Journal of Art Therapy* 14 29–37

Hogan, Susan & Pink, Sarah 2012 Visualising interior worlds: interdisciplinary routes to knowing. *in:* Pink, S (ed.) *Advances in visual methodology.* SAGE, London 230–248

Ikemi, Akira; Yano, Kie; Miyake, Maki & Matsuoka, Shigeyuki 2007 Experiential collage work: exploring meaning in collage from a focusing-oriented perspective. *Journal of Japanese Clinical Psychology* 25 464–475

Ingold, Tim 2011a Introduction. *in:* Ingold, T (ed.) *Redrawing anthropology: Materials, movements, lines.* Ashgate, Farnham 1–20

Ingold, Tim 2011b *The perception of the environment: essays on livelihood, dwelling and skill* Routledge, London

Manning, Erin 2009 *Relationscapes: movement, art, philosophy* MIT, Cambridge, MA; London

McCormack, Derek P 2008b Thinking-spaces for research-creation. *Inflexions* 1 1–15

McCormack, Derek 2012 Geography and abstraction: Towards an affirmative critique. *Progress in Human Geography* 36 715–734

O'Neill, Maggie 2008 Transnational refugees: The transformative role of art? *Forum Qualitative Sozialforschung* 9 article 59

O'Sullivan, Simon 2001 The aesthetics of affect: Thinking art beyond representation. *Angelaki – Journal of the Theoretical Humanities* 6 125–135

Pink, Sarah 2012 Advances in visual methodology: an introduction. *in:* Pink, S (ed.) *Advances in visual methodology.* SAGE, London 3–17

Pink, Sarah, Hogan, Susan & Bird, Jamie 2011 Intersections and inroads: Art therapy's contribution to visual methods. *International Journal of Art Therapy: Inscape* 16 14–19

Rappaport, Laury 1998 Focusing and art therapy: tools for working through post-traumatic stress disorder. *Focusing Folio* 17 http://www.focusing.org/arts_therapy.html. Accessed 05 Feb 2014.

Sullivan, Graeme 2010 *Art practice as research: inquiry in the visual arts* SAGE, London
Thornes, John E 2004 The visual turn and geography (Response to Rose 2003 intervention). *Antipode* 36 787–794
Tolia-Kelly, Divya P 2007 Fear in paradise: the affective registers of the English Lake District landscape revisited. *Senses and Society* 2 329–352
Tuan, Yi-Fu 1975 Place – experiential perspective. *Geographical Review* 65 151–165

CHAPTER 5

Progressions

Abstract Chapter 5 focuses on Gendlin's notions of progression and crossing in relation to verbal and visual concepts. Exploring progression as a coming into being of new representational forms, Banfield develops recent geographical efforts to rethink abstraction as productive rather than reductive through Gendlin's work. Considering progression as pre-reflective connectivity between established concepts, she addresses the potential to think more-than-logically about verbal concepts through a discussion of disciplinary efforts to eradicate terms with scalar associations, and about visual concepts in relation to artworks. Banfield proposes that with the crossing of multiple concepts operating in metaphorical fashion, the more-than-logical connectivity between and beyond the concepts themselves can be opened up to enable the explication (articulation) of new conceptual terms from that connectivity.

INTRODUCTION

In this chapter, I address Gendlin's thinking in relation to our ability to think with and from progressions between concepts. Gendlin proposes that statements can make mere logical sense or they can lift out more than the merely logical, stating that if we can follow the next step even if it does not follow logically, then it must follow in more-than-logical fashion: "It moves from what was 'more than logical'—from the 'lifted out'. That can be seen only in progressions." (Gendlin 1989: 406)

© The Author(s) 2016   81
J. Banfield, *Geography Meets Gendlin*,
DOI 10.1057/978-1-137-60440-8_5

With progression, Gendlin asserts, we do not lose logical power, but we do find that there is more specificity and precision in progressing from or between concepts than logic alone permits (Gendlin 1989). While statements can be considered in strictly linguistic terms, if we maintain Gendlin's open approach to concepts as including both visual and verbal, we can appreciate that these more-than-logical progressions are discernible not only between multiple verbal concepts but also between multiple visual concepts, and between a combination of the two. We can also think about progression as applying not just between two or more pre-existing concepts, but also between existing and emergent concepts, or in terms of the coming into being of a new concept, verbal or visual. There is a more-than-logical space beyond, between and prior to concepts, from which an emergent concept is or can be lifted out.

To interrogate these issues in this chapter, I draw on both my own field research with participating artists—specifically the empirical materials presented and discussed in Chaps. 3 and 4—and my own artistic practice conducted separately from the field research. In the first instance I work through the idea of progression as a process of abstraction, a lifting out of new meaning from implicit understanding. I engage with specific geographical work on the idea of abstraction to tease out the more-than-logical space between verbal and visual conceptualizations of my empirical material, and to consider the idea of progressions between concepts as a process of emergence, not just as a source from which emergence happens. Subsequently, I explore the idea of progression as a more-than-logical space between concepts through Gendlin's notion of the crossing of concepts, within which this implicit space becomes accessible. I consider two sets of pre-existing concepts through which to explore the idea of the more-than-logical progressions between them: I discuss geographical debates over the need to eradicate from geography's lexicon scalar terminology that holds inherent spatial or visual associations as an example of verbal concepts; and I draw on a selection of my own paintings to explore the more-than-logical space between them as an example of visual concepts. Through these discussions I suggest that Gendlin's conceptual work has much to offer geographical understandings of abstraction, with progression considered both as a more-than-logical space between, before and beyond concepts, and as a generic term for the emergence of concepts, which his specific explicatory techniques—to be addressed in Part 3—are designed to support and facilitate.

ABSTRACTION

The nature and function of abstraction is attracting renewed and critical attention in geography at present, with disciplinary tendencies being rethought and reframed. I focus particularly here on Derek McCormack's recent work as this provides a clear illustration of this rethinking (Banfield 2014). In a powerful but succinct critique, McCormack asserts that established tendencies to cast abstraction as a generalization and simplification, reduce the heterogeneity of lived embodied experience and are tarnished by associations with epistemological withdrawal from the world, objectification of the body, idealism and alienation (McCormack 2008b, 2012). Instead, abstraction is presented as a transformational process that participates both in the worlds that we inhabit and our efforts to make sense of those worlds. Such reformulations of abstraction have been discussed with specific reference to both the diagram, or line drawing, and the map, the archetypal geographical abstraction. As a particular example of abstraction, and a key abstraction within geography, diagrams have been defended against criticism of being reductive, detached and unable to capture the dynamism and complexity of experiential space and lived experience (McCormack 2008a, 2012); while maps have also been reconsidered as lived abstractions that act in the world, both adding to and transforming the world (Kitchin and Dodge 2007; Gerlach 2013).

Sketches, drawings and paintings are also forms of abstraction familiar to the practice of geography, and the "had" space-times crafted into being in the practices of participating artists discussed in Chaps. 3 and 4 support this affirmative understanding of abstraction. The juxtaposition of spatialities, temporalities and subjectivities challenges any assumed opposition between the lived and the abstract (Daniels 1985; Cresswell 2006; McCormack 2008b) and these "had" space-times were described as being implyingly more than occurred. My intention is to consider how the relation of symbolism and narrative in these artistic practices can inform geographical understandings of abstraction, by revisiting and reworking the narrative content of the empirical material presented in the earlier chapters in diagrammatic form, and by drawing into productive conversation the discussion of crossings and progressions in relation to artistic and linguistic explication of the implicit from the previous chapter, with McCormack's affirmative critique of abstraction (Banfield 2014).

Given the proposal that the geographies of practice and performance can be apprehended through the concept of the diagram (McCormack 2005),

I have reworked the narrative content of the empirical case study presented here in diagrammatic form, so that the text and image function as narrative and symbolism (Banfield 2014). To avoid associations of diagramming with either maps or formal schematic representations that might be found in textbooks, these particular iconographic renderings are best considered as glyphs, simple figures of lines, blobs, points and so on. Glyphs can be considered distinct from: icons, which work on the basis of resemblance; indices, which connote qualities; and symbols, which rely on conventions. By contrast, glyphs derive meaning from their gestalt properties and context, expressing concepts not easily conveyed by likeness, and encouraging generalization through their abstract nature (Tversky 2011; Banfield 2014). In this sense, the importance of gestalt properties is reminiscent of Gendlin's emphasis on the whole; context might be considered equivalent to interaffecting; and the expression of concepts not easily conveyed by likeness is suggestive of the non-specificity of the implicit.

In the empirical case study discussed in Chaps. 3 and 4, artistic subjectivity and spatiality emerged through two reciprocal relations of interrogation: the first between the artist and the subject through coordinated differentiation and felt sense; and the second between the artist and the viewer through the manipulation of narrative intelligibility. In Fig. 5.1, these are the loops to left and right. The point in the centre could be likened to the idea of occurrings as fulcra of interrogation, the resultant implyings of which might interaffect either or both of the relations of interrogation.

The spatiality of this paper tells the story of an artist embarking on their artistic practice (at either the top or bottom of the "S"), as constituted by previous experiences, but re-constituted differently through their practice as they move along the "S". On the way, the artist might alternate to varying degrees between the reciprocal relations of interrogation, might

Fig. 5.1   Empirical "had" space-time (*Source*: Author (2014))

favour one over the other, or might loiter at the junction between the two (Banfield 2014).

Although superficially this glyph is static, generalized and simplified, it does not isolate, extract or freeze movement. Instead it invites movement: traversing, circulating, reverberating and oscillating. It is both narrative and symbol, form and movement. It conjures movement even though it does not itself move. This invitation to mobility lies at the heart of at least three interrelated ways in which these glyphs challenge notions of the abstract as reductive, universal and repressive. Firstly, this diagram does not work against an understanding of lived experience (McCormack 2008a), but invites the revisiting and retelling of that lived experience. Secondly, it is generative, helping us to think again about the experience that it narrates (McCormack 2004, 2012). Thirdly, as it is never present in position but only in passing (Massumi 2002)—in its retelling—it is onto-genetic (Banfield 2014).

The glyph functions as two types of abstraction (McCormack 2008b, 2012). It is identifiable as a set of lines on a page, it also evokes the movement of thought in relation to the topic or content to which it speaks. While I would not deny that the first type of abstraction can inhibit the second, as McCormack (2008b) argues, I would additionally argue that the second can inhibit the first. Just as pausing in one's thoughts to jot ideas down on paper, in either discursive or figural form, can interrupt ongoing movements of thought, so too can ongoing movements of thought interfere with the jotting down of ideas on paper, as additional associations and developments that occur before the preceding ideas have been recorded can cause the preceding ideas to become lost, confused or diluted. Considering these issues with Gendlin's philosophy in mind, the former could be considered an example of a stopped process, which might or might not resume as previously implied following the pause, while the latter could be considered an example of an implying implying (or stimulating) its own change. Equally, though, pausing to jot down ideas is its own occurring, implying further or different movements of thought, which might or might not change what was previously implied, culminating in a complex interaffecting, generative of new thought and action (Banfield 2014).

I would also argue that the first type of abstraction can both exemplify and modify the second, as in the case of these glyphs. For Gendlin, imagery is a special kind of bodily living rather than a representation (Gendlin 1981), challenging identified criticisms of abstraction for elevating the visual logical of the rational mind (McCormack 2012). Explication is never representation, but always a further process (Gendlin 2001). As bodily change is involved in the making of the image from the movement of thought, we can ask two questions to illustrate the inherent connection between the two types of abstraction. First, we can ask what bodily changes does the movement of thought bring about in producing the diagram, image or glyph? Second, we can ask what changed bodily living is now implied consequent to the production of the image (Gendlin 1981), where bodily living includes its movement of thought? The movement of thought implies the production or occurring of a diagram or image, but the occurring of the image might change the implying, so that the subsequent bodily change might not be that which was originally implied. Further, the movement of thought occurs into the drawing of an image as bodily living, which is changed by the occurring, and which subsequently changes the movement of thought that initially implied the drawing (Banfield 2014).

With these two types of abstraction so tightly bound in the implying–occurring–implying cycle, the necessity for and utility of such a distinction is brought into question, at least in relation to diagrammatic abstractions. This distinction (McCormack 2012) seems to suggest that abstractions can be of one type (lines on a page) or the other type (the movement of thought), but not both at the same time, in the manner that I argue is the case here. Where abstraction as the movement of thought pertains to, for example, dance, this type of abstraction can indeed exist without any need for or connection with a paginated linear abstraction in the form of a diagram. However, the reverse does not hold for diagrammatic abstractions. Where abstraction is diagrammatic, a Gendlinian perspective suggests that the former type of abstraction (lines on a page) not only can exemplify the latter type of abstraction (movement of thought), but also that it must do so. This comparative consideration of the applicability of a differentiated conception of abstraction between dance and diagram is fully consistent with Gendlin's ideas concerning the relation between implicit and explicit understanding, in which the explicit is yoked to the implicit but not vice versa. This is not to say that we can think of the former type of abstraction (lines on a page) as explicit, and the latter type of abstraction (movement of thought) as implicit, as an image might be based entirely on implicit rather than explicit understanding, and dance might draw on concepts as well as or rather than affects. It is simply to say that abstraction as the movement of thought does not necessarily entail the drawing of lines on a page (although this is not precluded), but that abstraction as image or diagram does necessarily entail the movement of thought. Where diagrammatic abstractions are concerned, for practical or analytical reasons we might choose to disregard the inherent connection between the two types of abstraction, but that is an academic distinction that does not appear to be supported by the *Process Model* (Gendlin 1981, 2001; Banfield 2014).

Such glyphs are also generative, taking on a life of their own in helping to move our thoughts on, functioning as sketches that clarify and develop thought by encouraging a multitude of re/interpretations (Tversky 2011). Presenting the narrative content of my empirical material in other than narrative form provides a different means of engaging with that narrative. In essence, we come to think with abstraction rather than merely thinking about abstraction (McCormack 2012). In this light, the abstraction itself becomes a thinking-space (McCormack 2008b), shifting from retrospective awareness of an abstract form to a concurrent awareness of abstractive potentiality, and we see the inkling of a more-than-logical space around the abstract form, from which progression might proceed. Described as both a processual movement of thought and a privileged site at which this movement is amplified by novel configurations of things, ideas and bodies, the thinking-space is both a facilitative environment and a generative activity (McCormack 2008b). It constitutes Gendlin's progressions in both senses: as a more-than-logical space between, before or beyond concepts/abstractions; and as the process of conceptual emergence from more-than-logical, or implicit, understanding. The emergence of a "had" space-time in artistic practice is illustrative of such a thinking-space, and so too is the glyph presented earlier (Fig. 5.1). The glyph draws attention to notions of repetition, transformation, circulation, stasis, progression and alternation. These ideas are not explicitly drawn out in the narrative material itself, but neither are they absent from the narrative. These features of lived experience in artistic spatializing and subjectifying practices are more readily foregrounded in the glyph than in the narrative. By contrast, the specific details of interrogative processes and spatializing catalysts are foregrounded in the narrative but not in the glyph. Consequently, the glyph is not simply a reduction of the narrative, as its occurrings imply different but connected meanings (Banfield 2014). They cross with each other, in the Gendlinian sense that they mean more when considered in their crossing than they do when considered in isolation. Such amplification through novel configurations of ideas also constitutes their crossing as a thinking-space.

Interestingly, despite these differences, the glyph and the narrative mirror each other as each challenges commonly assumed oppositions. In the empirical material presented in Chaps. 3 and 4, dualisms such as real–ideal and narrative–symbolic were dissolved, while in the glyph repetition sits alongside progression, and transformation partners with stasis. Whether in narrative or diagrammatic form, and irrespective of the differences in the detail presented in each, we are encouraged to attend to coherence, interpreting conjointly that which would usually be described in mutually contradictory terms (Stengers 2008; Banfield 2014). By considering different abstractions in their crossing, we can draw out elements of the world to make them thinkable and sense-able (McCormack 2012). We are able to think and sense more of our implicit understanding in the crossing of concepts, than if we consider each concept in isolation, so the crossing, too, is both a "had" space-time and a thinking-space.

The diagram or glyph, then, holds together the abstract but real organization of forces without capturing them in specific subject–object form (McCormack 2005) and relates them in a visual form that is more-than-visual. In thinking with diagrammatic abstraction in this way, we are encouraged to reconsider and retell the narrative to which it relates with the diagrammatic emphasis on coherence in mind. Diagrammatic abstractions are productive, adding to the world as an operative force in our understanding of it, implicitly interaffecting our ongoing movement of thought in a facilitative rather than deterministic fashion (McCormack 2005; Toscano 2008; Gerlach 2013). This brings us to abstractions and diagrams as potentiality (Banfield 2014). In making complex materials available for manipulation, abstractions constitute an ontological transformation within, rather than a removal from, the world (McCormack 2012). As indicated above, the diagram enables us to revisit the narrative afresh, whether in textual or visual form, and transform our reading and telling of it. With the potential for constant recomposition, the glyph functions as a relational model rather than an exact copy of the narrative (Rees 1973).

The diagram in effect occupies its field of relational potential (Massumi 2002), in which the explicit lines of the diagram imply other and different occurrings, rich in opportunities for progression between these varied tellings. This is reminiscent of the artistic "had" space-times discussed in Chaps. 3 and 4, which were implyingly more than that which occurred in their finished representational form, reminding us that everything could always be other than it is. The mappings of meanings from that which is represented to the representing glyph are partial and variable (Tversky 2011), such that any individual telling of the narrative, and any individual experience of the artistic practices described therein, constitutes a re-drawing within that implied potentiality. Each may have any or all of the elements of the narrative encapsulated in the potential of the diagram, but the specifics of each instance are particular to their individual telling (Banfield 2014). Figure 5.2 presents this in glyphic form for the empirical material presented earlier, indicating a range of potential alternative tellings or "had" space-times.

The abstraction here has no ontological security. It only appears ontologically secure because it is reaffirmed in its repeated performance (Kitchin et al. 2013) in a co-constitutive production between the inscription, the individual and the world (Kitchin and Dodge 2007). This inscription takes two forms, the narrative and the iconographic, but in its co-constitution it too modulates how the world is understood (Kitchin and Dodge 2007), and in its productivity modulates how the world is. Reminiscent of the productivity of emergent "had" space-times in the empirical case study, the abstraction—narrative or diagrammatic—captures something of the world, while simultaneously doing work in the world; preceding and producing the territory that it purportedly represents (Kitchin et al. 2013). Abstraction here is ontogenetic, not ontological, unleashing its potential to be other than it is (Massumi 2002; Banfield 2014).

In its productive force, the diagram is generative of a space of enactment (McCormack 2005), and that which goes into the constitution of the space is determinant with respect to its diagrammatic (Jones et al. 2007). The diagram and the spatiality are mutually constitutive but not coterminous, as the spatiality is implyingly more than its diagrammatic occurring (Banfield 2014). The site does not precede but emerges from the interactive processes that assemble it (Collinge 2006; Escobar 2007), consistent with the emergence of artistic "had" space-times. Through this process, the site or artistic "had" space-time becomes coordinatedly differentiated as the movement of thought in the drawing of lines or marks on a page

Fig. 5.2   Alternative empirical "had" space-times (*Source*: Author (2014))

brings about changes in bodily living, which occur into the ongoing artistic practice process (Banfield 2014). A Gendlinian understanding of abstraction emphasizes the caution with which we should assume distinctions between different types of abstraction, and reinforces more affirmative understandings of abstraction as generative and bodily practices.

On this reading, then, glyphs potentially allow us to understand and work with geographical concerns in ways that are sensitive to the materialities and spatialities of the practice through which they emerged, because the interaction of implying and occurring brings about changes in both the movement of thought and bodily living, allowing for a multiplicity of emergent spatialities. They bring the possibility to work without recourse to metaphors, such as the container, grid coordinates, the network, the assemblage or the fold, all of which have certain visual elements inherent within them. Glyphs can provide for multiplicity, complexity, diversity and transitivity of spatialities, without constraining those spatialities to any pre-established visual or topological assumptions. On this Gendlinian reading, glyphs both allow and demand the carving of their own spatiality and visuality (Banfield 2014).

Importantly, neither the narrative nor the iconography need to be elaborate or sophisticated. While the narrative of the empirical material presented here, and the lived experiences of the artists described therein, might seem complex, the glyphs certainly are not. This introduces a seeming discrepancy between simplicity and complexity. However, the connectivity between the imagery and the narrative is crucial to dissolving this discrepancy. The imagery or narrative can be simple or incomplete while relating to complexity and dynamism in its counterpart. Equally, the simplicity and incompleteness of either the narrative or the imagery can contribute to complexity and dynamism in its totality, through the partiality of the mapping of meaning between that which is represented and that in which it is represented (Tversky 2011), providing scope and flexibility for alternative imaginings (Banfield 2014). It is in the crossing that the fullest implicit meaning can be found, and it is to crossings that my attention now turns.

CROSSING

In Gendlin's own writings, this idea of speaking from the progressions between pre-existing concepts gains specificity in his technique of crossing. For Gendlin, each of two crossed concepts means more in their crossing than they do in isolation (Gendlin 1995); each takes on more meaning in the context of the other than when simply taken on its own terms, functioning as an implicit metaphor. In relation to the crossing of concepts, which can also apply to the crossing of emergent explication of the implicit in their as yet unformalized forms, Gendlin's focus is primarily on linguistic explication. In this section, I consider the crossing of verbal or linguistic concepts in relation to disciplinary debates around scalar terminology, before exploring the crossing of visual or artistic concepts in relation to two of my own paintings.

Linguistic Crossing

Contemporary geography is equipped with an arsenal of spatial concepts and metaphors, including network, mesh, fluid, rhizome and assemblage (Hetherington and Law 2000; Simonsen 2004; McFarlane 2009; Knappet 2011), which have progressed from hard and fixed to porous, blurred and liquid imaginaries (Sheller 2004). By the early 1990s scale had been described as, arguably, geography's core concept, and as theoretically and empirically problematic (Herod 1991), with calls arising for greater clarity and new ideas in relation to this unacceptably vague concept (Howitt 1993). Similar pleas were heard at the turn of the millennium, when particular spatial metaphors were deemed to be becoming overly dominant and new conceptual terminology was sought that avoided implied ontological or spatial fixity (Hetherington and Law 2000). By the middle of the first decade of the twenty-first century, the flurry of conceptual motifs (Lorimer 2007) that had sprung up included both hierarchical and networked concepts, but among diverse disciplinary perspectives a growing number of theorists remained dissatisfied with dominant understandings of scale (Marston et al. 2005).

While some authors advocated flat ontologies or topological thinking, rather than scalar thinking, on the grounds that this would avoid problematic distinctions between small and large, or abstract and concrete (Amin 2002; Marston et al. 2005; McNeill 2010), others argued that scalar concepts were so prevalent that it was impossible to think space without them and any attempt to refuse them would be fruitless (Jones et al. 2007; McFarlane 2009). Flat ontologies were not universally popular as alternative spatial imaginaries, either, with some authors arguing against flatness for its inability to visualize relationship strength (Inkpen et al. 2007), and others preferring a straight critique of existing concepts to the generation of a plethora of new candidates (Hoefle 2006). Even among advocates of flat ontologies, no single approach was consistently articulated. Some were amenable to retaining an element of scalar reasoning but allowing the development of alternative conceptual understandings (Amin 2002), while others argued for the outright expurgation and replacement of scalar terms (Marston et al. 2005).

Gendlin's writings offer potential with regard to several perspectives on scalar imaginaries. On the one hand, his explicatory techniques make possible the outright replacement of existing vocabularies by generating new concepts from the implicit. On the other hand, those same explicatory techniques also allow for the implicit refinement and development of existing vocabularies to liberate them from supposedly problematic associations. In particular, thinking-at-the-edge is consistent with Collinge's (2006) argument that, rather than eliminating terms such as scale from the discipline's lexicon, geographers should work their metaphysical terminology back against itself to reinscribe it into the context from which it came. For Gendlin the context from which the terminology came is our implicit understanding, and the task is not so much one of reworking our terminology back against itself, than one of reworking our terminology back through the implicit (Banfield 2014). Further, Gendlin's notion of crossing provides an alternative approach to conceptual development. In the absence of any deliberate attempts to access the implicit, the crossing of two concepts is similar to conceptual development through comparative integration of multiple concepts, but, in combination with deliberate attempts at accessing the implicit, the crossing of concepts might offer additional benefit in developing new concepts from the more-than-logical that lies between them. The previous discussion of progressions between narrative and symbolic abstractions also suggests that such radical expurgation is not necessary. Both narrative and diagrammatic practices might productively avoid association with any particular spatial imaginary, delivering the desired n diagrams for specific and singular sites (Marston et al. 2005), while retaining scalar terminology for occasions when its employment is appropriate.

Three potential responses to disciplinary attempts to link hierarchical and networked concepts have been proposed: (1) to supplement hierarchical concepts with network features; (2) to develop hybrid forms; or (3) to replace the conceptual system entirely by collapsing one into the other (Marston et al. 2005). For the purposes of this discussion, I use this triad of possibilities to explore what Gendlin might mean by the crossing of concepts within an explicitly geographical conceptual arena. As a route into that discussion I begin by considering briefly the capacity for crossing to function as a form of critique consistent with Hoefle's (2006) stated preference in this direction.

If, in their crossing, two concepts mean more than they do individually, the very practice of crossing acts as a simultaneous critique of both independent terms, allowing for the identification of features in each that are potentially amenable to modification to enable their incorporation into a formalized crossed concept. To a certain extent, and on a conceptual level, this is one way in which theoretical development already takes place, and can be likened to the idea of hybrid concepts. However, with Gendlin's idea of crossing, it is not merely that pre-existing conceptual units are crossed but that in their crossing, new more-than-logical space is opened up, from which new conceptual knowledge can be explicated or lifted out. This introduces a distinction between hybrid and crossed concepts, as a crossed concept is not developed by a straightforward hyphenation of two pre-existing terms, and it does not simply refer to elements of each of the pre-existing concepts. Instead, a crossed concept is generated as a new conceptual unit which, through interaffecting, is implicitly rather than explicitly connected to the original concepts that were crossed. Potentially, this responds to two particular criticisms of existing assumptions about categorical definitions which plague debates concerning appropriate spatial imaginaries: that they are assumed to be logically distinct; and that their boundaries are assumed to be rigid (Howitt 1993). Both the logical distinctiveness and the rigidity of their boundaries are rendered questionable within a Gendlinian understanding of interaffecting and implicit–explicit connectivity.

Although we can fairly readily integrate concepts on a conceptual or logical level through geography's penchant for hyphenation, how this reworking might be accomplished on an implicit level in a practical sense is a different matter. Looking at the horizontal and vertical spatial associations listed by Marston et al. (2005: 420), we could draw on Gendlin's idea of crossing in a number of ways. For example, we could consider how concepts of horizontality and verticality might cross with each other and how this might impact on our understanding of the terms listed; we might select a particular term as a conceptual fulcrum through which to bring the two together; or we might select a particular term from each category and cross them. As a case in point, layered is identified as a vertical geography, but it is arguably the plurality of layers rather than the notion of layer that evokes verticality. Such complications in our conventional or rule of thumb understandings have the potential to function as interrogative prompts to go beyond conceptual categorizations and consider our implicit sense of layer, verticality and horizontality, prompts that multiply if we cross layered (specified as a vertical geography) with dispersed (specified as a horizontal geography). For Gendlin, this is not about thinking about the terms as logical constructs, visual orientations or geometric axes, but to move from those understandings to our implicit understanding, to gain a felt sense of horizontality and verticality.

Thinking across concepts logically might, given their underlying inter-affecting, make available our implicit understanding of this interaffecting even without our awareness of it, but to maximize the implicit grounding within which emergent crossed concepts are rooted, the application of Gendlin's explicatory techniques to the crossing process might be particularly beneficial. By providing us with a connection between our conceptual understanding and our implicit understanding of new socio-spatial dynamics, such methods might enhance the sharpness of the revised definitions in both senses of the term, helping us to overcome the redundancy that afflicts old definitions of scale in the context of new socio-spatial dynamics (Howitt 1993). Concepts developed in such a way could then be employed either to supplement hierarchical imaginaries, as hybrid elements, or as replacements for existing imaginaries, by providing a new suite of spatial concepts, which is neither hierarchical nor networked but which implicitly retains aspects of both in a non-conflictual fashion. My own attempts at implementing some of these explicatory techniques, their potential, pitfalls and implications, are addressed in Part Three. For now, I leave an invitation with you to explore the potential of Gendlin's ideas of crossing concepts and progressions between concepts for geography's disciplinary debates around spatial terminology and imaginaries.

Artistic Crossing

Gendlin's ideas of crossing and progression perhaps provide a supportive scaffold for the development of new spatial understandings of a linguistic kind, but these same ideas can also be considered in relation to visual or artistic concepts. As previously observed in relation to art exhibitions, if the communicative effects and affects of an exhibition spring not from the individual artworks but the spaces between them, new aesthetic configurations can be forged on a larger, even transnational, scale (McNeill 2010).

Although the original context for this proposition was the potential for aesthetic connectivity between remotely exhibited artworks, it clearly resonates with Gendlin's ideas. Aesthetic significance emerges from the more-than-logical connectivity between, rather than the conceptual contents of, the individual artworks. In their crossing, the works exhibited mean more than they do independently. The works might have been produced by different practitioners and might have little or nothing in common in terms of conceptual content: their connectivity is not conceptual yet there exists an implicit connectivity, a more-than-logical progression, between them, from which we can lift out new meaning.

In order to explore how the communicative effects and affects manifested in the spaces (progressions) between crossed works allows for the lifting out of new meaning in the context of visual or artistic concepts, I draw on my own hobby artistic practice. I orient this discussion around two paintings, presented in Figs. 5.3 and 5.4, which explore artistically my first year's experience of keeping chickens. Figure 5.3 is more straightforwardly representational and is more immediately recognizable as a chicken than Fig. 5.4. Although the structural features of the two paintings and the colour palette employed are similar, they are also distinctive. Figure 5.3 is not a portrait of any particular chicken, but it is a depiction of a chicken in the style of a portrait. It is very much about the appearance of a chicken, in terms of form, colour and texture.

Fig. 5.3 Chicken Oil pastel and inktense (*Source*: Author (2012))

Fig. 5.4   Inner Chicken Oil pastel and inktense (*Source*: Author (2012))

By contrast, Fig. 5.4 is more of an evocation than a depiction, and its subject matter is the experience of drawing (gutting) a chicken for the first time; its felt sense. While form, colour and texture are evoked in visual form, it is the experience of the gutting process, rather than the appearance of the entrails, that forms the subject matter of the work. The egg that was about to be laid when the bird was slaughtered, the gizzard, the intestines and the trail of embryonic eggs are all there, but they are not depicted in an anatomically accurate manner in the way that the eye, the comb and wattle are in Fig. 5.3.

Adopting a comparative basis is one way in which we can look across the two paintings, but this is not my sense of what Gendlin means by the progression between concepts. The significance of these two paintings together and their relation to each other can be considered from a Gendlinian perspective in terms of both the finished works as paintings and the processes of their production.

In thinking about these paintings in terms of progression between them, we might be better off not identifying the difference in anatomical correctness between them, but in attending instead to the implicit meaning of the almost-laid egg in Fig. 5.4, in the location of the eye in Fig. 5.3, and similarly for the cluster of eggs in the position of the wattle. We perhaps detect implicit awareness of the concurrent ephemerality and perpetuity of life; the life-giving capacity of a hen in a position anatomically associated with giving access to the inner life of a creature (the eye); and in the position of an anatomical feature which indicates the particular life stage of the hen as mature enough to lay (the wattle). We can associate this with Gendlin's implying–occurring–implying cycle, as the implying and occurring of the digestive process gives rise to the occurring of bodily form and growth. Finally, although the beak in Fig. 5.3 is mirrored by a similar structural feature in Fig. 5.4, it is neither the same nor is it different. While not being directly representational, the sense of flow in Fig. 5.4 from top right to bottom left parallels that of Fig. 5.3. Although there is no clear anatomical structure in Fig. 5.4, the movement of material through the chicken from beak to vent is consistent between the two. This is implicitly connected to the moral and ethical issues with which I grappled in keeping chickens for meat as well as for eggs. I had been feeding the chickens so that they could feed me. For the chickens to feed me I had to kill them. By killing the chickens I gained the meat but lost the eggs. The flow of food, energy and other matter through the respective bodies of human and chicken in their body-environment concretion was palpably experienced in the process of preparing the birds for the table, and my conceptual wrangling with those moral and ethical issues was closely connected to my sense of being part of an implicit many with the chickens, which sub-served my moral and ethical concerns. This is not to suggest that these potential crossings were consciously or deliberately incorporated into the paintings, nor that these meanings are inherent within the paintings ripe for interpretation, but that in considering the two paintings together, further implicit meaning can be made available for explication than would be the case for either in isolation.

In the sense of thinking-at-the-edge, the representational contents of the works in their crossing afford access to the implicit understanding informing them. The important point emerging from this discussion is that none of these aspects would have been apparent if we considered only one or other of the paintings in isolation. It is in the progression between them when they are crossed that further implicit meaning becomes available for explication.

This understanding is reinforced if we consider the process of production of these two paintings. Figure 5.4 was produced before Fig. 5.3, and was never intended to be a representational depiction of a chicken, from either inside or outside the chicken, but was simply intended to evoke the colours, textures and sensations of a particular experience; its felt sense. The fact that Fig. 5.3 depicts a chicken in the orientation and close-up fashion that it does was therefore implied by the occurring of Fig. 5.4, particularly in terms of its structure, rather than the other way around. The almost-laid egg and the gizzard were not depicted in the position of the eye, but the eye was depicted in the position of the egg and gizzard. The occurring of Fig. 5.4 implied the occurring of Fig. 5.3, and once both have been seen together, the occurring of Fig. 5.3 implies a different understanding (or occurring) of Fig. 5.4.

Another type of occurring can be seen here, as it was not me who identified the potential for the structure of Fig. 5.4 to accommodate the anatomical form of a chicken for Fig. 5.3. Credit is due to my parents for that, who were trying to determine what I was painting as it unfolded. When Fig. 5.4 was partially complete, my father suggested that it might be a chicken, which was correct, but he meant it as a portrait of a hen rather than an evocation of entrails. I could not see a chicken in the painting at that stage, but once he indicated where the eye, beak and so on were perceived to be, so too could I imply the painting to be a portrait of a chicken. Notably, the implying of a chicken portrait did not lead to the occurring of Fig. 5.4 as a portrait. That painting became less portrait-like as it progressed, because the structural indicators became less prominent as the colours and experiences developed. However, the implying of that painting in portrait form—although not occurring as one—also implied the occurring of Fig. 5.3 in portrait form.

Taking this discussion one step further, my father also made an alternative suggestion as to what the painting might be. Based on his viewing of it in portrait rather than landscape, he saw an abstract face (Fig. 5.5).

In terms of Gendlin's philosophy, the painting that was originally implied (the evocation of a particular experience) very much implied its own change. The occurring of my father's viewing of the work and articulated understanding of it changed (added to) the original implying. This implying has not occurred, in the sense that the painting that was produced was not produced in accordance with my father's implying of it. However, its implying is ongoing because the painting that did occur (Fig. 5.4) is still implied in that manner. This particular example brings out quite powerfully the notion of occurring into implying, and of implying stimulating or implying its own change, by virtue of the occurring that was implied changing that which was originally implied.

The crossing of, or progression between, concepts, then, can be applied to both visual and verbal concepts, and to the processes of their production and their finished form, recalling the earlier discussion of artistic practice as a process of conceptual sharpening. Crossing potentially enhances

Fig. 5.5 Abstract Face Oil pastel and inktense (*Source*: Author (2012))

our ability both to access and apprehend implicit understanding, as two concepts are implicitly more meaningful in their crossing than in isolation, suggesting further avenues for geographical engagement with Gendlin's work in relation to developing disciplinary capability to work conceptually with images (Hawkins 2015). It is worth drawing out a distinction here between crossing and progression. Earlier, I characterized Gendlin's idea of progression in two ways: as abstraction (explication) and as the more-than-logical. As I understand it, progression is both the lifting out (abstracting or explicating) and that from which it is lifted out (the more-than-logical). Crossing makes available or opens up progression between concepts, as that from which new meaning is lifted out to enable progression as the lifting out. While this dualistic nature of progression might seem reminiscent of Massumi's notion of affect, as both that which escapes capture and the connective means by which it is partially captured (Massumi 1995), it is worth remembering that Gendlin's progression between concepts is not the implicit in its entirety but a circumscribed "had" space-time between two or more concepts. Similarly, progression as lifting out is not a connective means but an active process, through which the circumscribed implicit meaning between concepts is lifted out into conceptual form. It is also worth remembering that Gendlin does not consider the implicit to escape in the way that Massumi considers affect to escape; the implicit is not captured but carried forwards (Massumi 1995; Gendlin 2009b). Rather than being problematic, then, I think of this dual understanding of progression as shorthand for specific instantiations of Gendlin's broader ideas of the implicit and explication.

CONCLUSION

Gendlin's concern with the relation between the explicit and implicit is not centred solely on the progressive emergence, or lifting out, of concepts from the implicit, but also on how multiple concepts relate to each other in more-than-logical ways, and our ability to enter and explicate from this more-than-logical connectivity. These two meanings of progression can perhaps be conceived as relating to Gendlin's two types of implying, whereby progression as the emergence of concepts from the implicit is consistent with temporal implying, and progression as implicit connectivity between concepts is consistent with horizontal implying. In both cases, the implying is ongoing, but we can occur into that implying, for example through focusing and thinking-at-the-edge, to explicate from the implicit.

In this chapter I have explored the temporal implying of the emergence of concepts in the context of geographical engagement with abstraction, and I have explored the horizontal implying between concepts in the context of both linguistic and artistic crossings. When applied to artistic or visual concepts, and particularly if applied in combination, these ideas and practices might provide a valuable contribution to emerging efforts, both to revisit critically and revise implicitly the host of concepts (scalar or otherwise) with which geography repeatedly wrestles (see, for example, Marston et al. 2005; Collinge 2006; Escobar 2007; Jones et al. 2007), and to enhance geography's capacity to think conceptually about images and image-making (Hawkins 2015). The key question that remains to be addressed, and which I seek to address in Part 3 is, how can we operationalize these ideas and Gendlin's psychotherapeutic techniques for accessing and explicating from implicit understanding within geographical research, and with what implications?

REFERENCES

Amin, Ash 2002 Spatialities of globalisation. *Environment and Planning A* 34 385–399
Banfield, Janet 2014 *Towards a non-representational geography of artistic practice.* Unpublished doctoral thesis, University of Oxford Forthcoming online: https://ora.ox.ac.uk:443/objects/uuid:dd12e1c4-f222-435b-adc0-c1bb68e4f4ac

Collinge, Chris 2006 Flat ontology and the deconstruction of scale: a response to Marston, Jones and Woodward. *Transactions of the Institute of British Geographers* 31 244–251

Cresswell, Tim 2006 *On the move: mobility in the modern western world.* Routledge, New York; London

Daniels, Stephen 1985 Arguments for a humanistic geography. *in:* Johnston, R J (ed.) *The future of geography.* Methuen and Co. Ltd, London 143–158

Escobar, Arturo 2007 The 'ontological turn' in social theory. A commentary on 'Human geography without scale', by Sallie Marston, John Paul Jones III and Keith Woodward. *Transactions of the Institute of British Geographers* 32 106–111

Gendlin, Eugene T 1981 Focusing and the development of creativity. *The Focusing Folio* 1 13–16 http://www.focusing.org/arts_therapy.html. Accessed 05 Feb 2014

Gendlin, Eugene T 1989 Phenomenology as non-logical steps. *in:* Kaelin, EF & Schrag, CO (eds.) *American Phenomenology: origins and developments.* Kluwer, Dordrecht 404–410

Gendlin, Eugene T 1995 Crossing and dipping: Some terms for approaching the interface between natural understanding and logical formulation. *Minds and Machines* 5 547–560

Gendlin, Eugene T 2001 *A Process Model* The Focusing Institute, New York

Gendlin, Eugene T 2009b We can think with the implicit, as well as with fully-formed concepts. *in:* Leidlmair, K (ed.) *After cognitivism: a reassessment of cognitive science and philosophy.* Springer, London, New York 147–161

Gerlach, Joe 2013 Lines, contours and legends: coordinates for vernacular mapping. *Progress in Human Geography* 1–18 doi: 10.1177/0309132513490594

Hawkins, Harriet 2015 Creative geographic methods: knowing, representing, intervening. On composing place and page. *Cultural Geographies* 22 247–268

Herod, Andrew 1991 The production of scale in United States labour relations. *Area* 23 82–88

Hetherington, Kevin & Law, John 2000 After networks. *Environment and Planning D: Society & Space* 18 127–132

Hoefle, Scott W 2006 Eliminating scale and killing the goose that laid the golden egg? *Transactions of the Institute of British Geographers* 31 238–243

Howitt, Richard 1993 'A world in a grain of sand': towards a reconceptualisation of geographical scale. *Australian Geographer* 24 33–44

Inkpen, Rob, Collier, Peter & Riley, Mark 2007 Topographic relations: Developing a heuristic device for conceptualising networked relations. *Area* 39 536–543

Jones III John P; Woodward, Keith & Marston, Sallie A 2007 Situating flatness. *Transactions of the Institute of British Geographers* 32 264–276

Kitchin, Rob & Dodge, Martin 2007 Rethinking maps. *Progress in Human Geography* 31 331–344

Kitchin, Rob; Gleeson, Justin & Dodge, Martin 2013 Unfolding mapping practices: a new epistemology for cartography. *Transactions of the Institute of British Geographers* 38 480–496

Knappet, Carl 2011 Networks of objects, meshworks of things. *in:* Ingold, T (ed.) *Redrawing anthropology: materials, movements, lines.* Ashgate, Farnham 45–64

Lorimer, Hayden 2007 Cultural geography: Wordly shapes, differently arranged. *Progress in Human Geography* 31 89–100

Marston, Sallie A; Jones III John P & Woodward, Keith 2005 Human geography without scale. *Transactions of the Institute of British Geographers* 30 416–432

Massumi, Brian 1995 The autonomy of affect. *Cultural Critique* 83–109

Massumi, Brian 2002 *Parables for the virtual: movement, affect, sensation* Duke University Press, Durham, NC; London

McCormack, Derek P 2004 Introduction: techniques and non-representation. *in:* Thrift, N & Whatmore, S (eds.) *Cultural geography: critical concepts in the social sciences, vol 2 practising culture.* Routledge, London 3–16

McCormack, Derek P 2005 Diagramming practice and performance. *Environment & Planning D: Society & Space* 23 119–147

McCormack, Derek P 2008a Geographies for moving bodies: Thinking, dancing, spaces. *Geography Compass* 2 1822–1836

McCormack, Derek P 2008b Thinking-spaces for research-creation. *Inflexions* 1 1–15

McCormack, Derek 2012 Geography and abstraction: Towards an affirmative critique. *Progress in Human Geography* 36 715–734

McFarlane, Colin 2009 Translocal assemblages: Space, power and social movements. *Geoforum* 40 561–567

McNeill, David 2010 Art without authors: Networks, assemblages and 'flat' ontology. *Third Text* 24 397–408

Rees, Ronald 1973 Geography and landscape painting: an introduction to a neglected field. *Scottish Geographical Magazine* 89 147–158

Sheller, Mimi 2004 Mobile publics: Beyond the network perspective. *Environment and Planning D: Society and Space* 22 39–52

Simonsen, Kirsten 2004 Networks, flows, and fluids – reimagining spatial analysis? *Environment and Planning A* 36 1333–1337

Stengers, Isabelle 2008 A constructivist reading of process and reality. *Theory, Culture and Society* 25 91–110

Toscano, Alberto 2008 The open secret of real abstraction. *Rethinking Marxism* 20 273–287

Tversky, Barbara 2011 Visualizing thought. *Topics in Cognitive Science* 3 499–535

PART III

Exploring Gendlin's Methods through Artistic Practice

CHAPTER 6

Explicating the Implicit

Abstract This chapter describes the implementation of interview methods derived from Eugene Gendlin's psychotherapeutic techniques to articulate affect in geographical research into artistic practices. Making a valuable contribution to the development of research methods to apprehend affect, Banfield highlights the potential for both linguistic improvisation and formal conceptual understanding to afford access to pre-reflective understanding. She discusses seemingly contradictory research findings, which are accounted for in the context of Gendlin's philosophical ideas, elaborating these ideas further. She also outlines the implications arising from this first geographical engagement with Gendlin's work for methodological invigoration within non-representational geography, highlighting the need for multi-stage, multi-method and individually-tailored research designs.

INTRODUCTION

In this third and final part, I move on to practical matters associated with Gendlin's specific psychotherapeutic techniques, which are designed to aid the emergence of new conceptual understanding from as yet unarticulated implicit understanding. This process, Gendlin's explication, is the 'lifting out' of concepts from implicit understanding, or the progressions between, beyond or before concepts.

© The Author(s) 2016
J. Banfield, *Geography Meets Gendlin*,
DOI 10.1057/978-1-137-60440-8_6

In this chapter I undertake two tasks. First, I detail my experimental application of explicatory techniques adapted from those of Gendlin in my field research with practising artists in order to suggest that such methods hold potential for geographical research into affect. The second is to consider seemingly contradictory implications of this research in the context of Gendlin's philosophical and methodological writings. Subsequently, I identify potential implications that arise from tying empirical outcomes of Gendlin's explicatory techniques into the philosophical framework from which they were derived, as a forerunner and springboard to a fuller critique of my engagement with Gendlin in the next chapter.

Experimenting with Explicatory Techniques

My explicatory methods were employed in the closing interviews, during which participants reviewed video footage of their artistic practice from the production sessions. Having attended a two-day masterclass on Gendlin's methods at the University of Bournemouth in the summer of 2011, I was familiar with the aims of Gendlin's techniques and the experience of their application, as well as the recommendation for a lengthy period of training on the part of both client and the analyst to maximize their effectiveness in therapeutic sessions. As I was unlikely to be able to deliver such lengthy training to participants or to undertake it myself, I drew on Gendlin's published and online guidance on the training necessary for the application of his methods. I also drew on Stelter's (2010) experience-based body-anchored interviewing technique, which builds on Gendlin's work but does not require extensive training, to develop a streamlined version of thinking-at-the-edge (Gendlin 2009a, b; The Focusing Institute). As thinking-at-the-edge seeks to strengthen the implicit meaning of an existing or emergent concept, I anticipated beginning with some form of conceptual content from the research sessions, perhaps an image, a word or an extract from an audio-video recording. I aimed to re-establish the felt sense of that conceptual content and support further articulation of its implicit meaning by whatever means felt most appropriate in the research setting. Either vocally or in writing, I asked participants to generate an initial articulation, which might be a grunt or a sigh, or it might be a proper word, a nonsense word or a stream of words. That articulation then became the focus of questions concerning whether that utterance, in whatever form it was made, captured the full implicit sense intended. A frequent approach was to ask, if that could mean just what you want it to mean, what would it mean? The most significant element of this re-description subsequently became the focus of the next attempt at thinking-at-the-edge, to push the limits of an utterance's implicit meaning and formal conceptualization. This process continued until the participant indicated that the articulation captured all of the implicit meaning necessary.

The closing interviews combined a streamlined Gendlinian interview style with visually stimulated recall (video-elicitation) of the practice-based sessions, and re-enactment interviewing (Drew 2006), which prompts re-enactment as a means of recreating experience. Participants were asked to try to connect with their implicit experience from the practice-based sessions while viewing their practice on screen—for example, by trying to evoke the materials, moods and senses of the occasion—and were encouraged to re-enact particular actions or behaviours if they felt that it would be beneficial in re-establishing the felt sense of their practice. I sought to maximize the potential for explication by bringing together multiple forms of intimate knowledge, such as real-time practice and social technologies, to increase participant access and alertness to their implicit understanding of their practice. The real-time artistic practice during the production sessions was intended to function as a form of focusing. The use of video-elicitation techniques aimed to utilize the telepresence of social technologies, affording further access to affective aspects of participants' experiences from the production sessions through the observation of their own actions on video (Gibbs 2009; Featherstone 2010; Blackman 2012). Re-enactment aimed to support this re-engagement with prior affective experience by recreating equivalent conditions of practice.

The modification and application of Gendlin-inspired explicatory techniques through these varied approaches sought both to cater for individual differences in affective re-engagement, and to compensate for the lack of participant training through the convergent employment of multiple means of accessing the implicit. Here, I draw primarily upon my work with two participants—Laura and Jane—whose closing interviews were particularly informative, especially when considered as a pair, before considering the success of these methods more critically in the following chapter.

During the initial viewing, I asked Laura to try to reconnect with her original experience of "engaging with your equipment and materials, the marks you're making, the effects you're seeking, and the sensations you're experiencing". Before the second viewing I asked her "if you had to give a name or word to the different ways in which you're engaging with your materials, what would they be?" In advance of the next viewing I asked

Laura to "note down in whatever form or format comes to you how you would characterize that clip". Although I changed the emphasis at this stage from words to formats, Laura's response remained linguistically rooted, and from the paragraph of text produced I asked Laura to identify three key words or phrases "which are the most significant to convey what you want to convey", before requesting that she expand on these (Banfield 2014). There was no evidence of linguistic improvisation, which from a Gendlinian perspective might suggest that Laura was articulating directly from her implicit understanding. However, repeated and alternating expansion and contraction of the material communicated did generate additional insight into Laura's implicit experiences during her artistic practice.

In relation to the first clip, in which Laura is seen preparing for a commissioned artwork on paper before starting to paint on canvas, words initially generated included "intuitive", "challenge", "connecting", "searching" and "smoothing". These were subsequently characterized as "to prepare for", "to get a feel for" the "colours, materials and also the composition", which was then elaborated further. Laura explained that working on paper before painting on canvas locks her into the feeling of the paints, which helps her to overcome her fear associated with working on canvas; an aspect of her artistic practice and experience which did not feature in her conventional interview account of her practice. In relation to the second clip, in which Laura is seen painting with both hands, I first requested a response to the clip as a whole, which generated "ambidextrous", "physical", "sculptural", "scrubbing" and "manipulating". On the next viewing, I asked Laura to isolate her experience of each hand individually, which generated (for the left hand) "control", "detail", "accuracy", "drawing", and (for the right hand) "scrubbing", "manipulating", "responding". Laura went on to explain that she was delineating detail with the left hand, with the right hand diffusing the sharper marks that the left hand makes. Particularly telling is that the work of Laura's left hand did not rate an explicit mention in her summary of the clip as a whole, even though her left hand is her dominant hand. This raises questions as to the relative degree to which left and right handedness might influence the effective explication of implicit understanding through artistic practice, especially in relation to the potential for linguistic explication subsequent to artistic explication. Along with the additional detail gleaned from the repeated expansion and contraction of meaning, it also alerts us to the myriad minutiae of practices that are not commonly captured in conventional retrospective interviews (Banfield 2014, 2016a).

In Jane's closing interview I selected a video clip that included a comment from Jane about the paintbrush enjoying itself. Before viewing the video, Jane clarified that "I like the gestural quality of it and I think that's allowing the brush to sort of enjoy itself". During the video review Jane elaborated further, saying that there is a risk in how hard to press on the brush because it is a combination of the texture, the brush, the paper and how damp everything is that is "letting an effect develop that you want so the brush is doing the work". As in Laura's account, Jane's emphasis is on physical and textural rather than visual aspects of her practice. Jane's comments associate the brush's enjoyment not only with a textural effect and the agentive force of the brush in generating that effect, but also with a gestural quality. This is not to equate the gesture with the mark on the paper or with a particular body form, but to consider this gestural quality in its unmappable tendency towards movement (Manning 2009). The gesture begins in advance of the initiation of the marking of the paper and continues beyond the break of contact between brush and paper; the mark that remains is but the artistic trace that bears witness to the passing of a gestural act (Ingold 2007; Banfield 2014).

Through her preparatory activity on paper, Laura gets a feel for her materials, the virtual force of their movements taking form, and channels that movement capacity into a bodily sensitivity and readiness, which is subsequently expressed in her work on canvas (Manning 2009). Jane's account describes how this sensitivity and responsiveness is expressed in her work, literally ex-pressed through her pressure on the brush. This pressure is intended to let the brush itself do the work through its own movement capacity, setting in motion the force of their taking form (Manning 2009). Together, Laura's and Jane's accounts illustrate the occurring of their actions into the implying of their materials; a body-environment concretion within which the artist focals these forces (as a bodily implying) into a singular interaffecting behavioural sequence.

Importantly, Jane continued "you can really scruffle it and, you know, you'd like to invent about fifty words, you know, of all the different textures that can be created". Jane's linguistic improvisation in generating the word scruffle, from a Gendlinian perspective, suggests that she was speaking from her implicit understanding, or felt sense, of that particular gestural texture. Beyond the aims of the original research, further work would be necessary to identify and describe the ways in which scruffle is meaningfully connected to other formal concepts, especially as the spelling is my own and if spelt differently—skruffle, scruphel, and so on—the word might connect in very different ways to different conceptual understandings. Further work would also be necessary to test the implicit intelligibility of this term to other people, and its translatability across different artistic media and traditions. However, this improvised scruffling hints at the potential productivity of techniques adapted from Gendlin's explicatory methods in apprehending affect, even in the absence of formal or lengthy training (Banfield 2014).

Of further note here is the seeming contradiction between Jane's assertion that it is the brush doing the work in generating the effect and her statement that she can really scruffle it, locating the source of control in her hand rather than the brush. Further, Jane's comment about the risk associated with pressing the brush down emphasizes the centrality of her hand in negotiating forcefulness and responsiveness, and echoes the negotiation evident in Laura's earlier account, at least in relation to the practice of her right hand. Recalling Jane's characterization of the brush enjoying itself as gestural, this centrality of the artist's hand is reminiscent of Tallis's discussion of gesture, in which prehension and comprehension are fused (Tallis 2003, 2004). Constantly iterated through the actions of the hand, the existential intuition [that] I am [this] is proposed as the moment of explicit indexicalization (Tallis 2004), by means of which we translate from implicit to explicit understanding. In this context, Gendlin's efforts at explication seek to facilitate the translation of implicit understanding through this indexicalization. The hand is the conduit between the implicit and the explicit, allowing for explication from implicit understanding, with gesture taking on both linguistic and artistic significance (Banfield 2014). However, the divergence in activity and experience between Laura's left and right hands suggests that the emphasis on *the* hand might need rethinking to account for differences in handedness and ambidextrous practices. Although beyond the scope of the current work, the issue of handedness might have implications for the effectiveness of explication and for the intersubjective intelligibility of explicated terms, which would benefit from further investigation.

Further emphasizing the importance of the hand in explication, when prompted to generate other such words, Jane responded that "the trouble is, I'd have to actually be doing [the] painting at the time" as "there are certain textures that you'd have to actually be doing I think to make the words", again suggesting that artistic doing might aid linguistic saying.

Interestingly, though, Jane had not generated linguistic novelties in the real-time production session, but only did so while viewing the video footage, and re-enactment failed to elicit any further linguistic improvisation. This indicates more complex relations between doing, viewing and saying than Jane's suggestion that doing will generate saying, and complicates debates concerning the benefits of real-time accounts of practices compared with qualitative interview accounts (Hitchings 2012; Banfield 2016a). Interviews and practice-based research might both be insufficient in isolation to capture activities and performances comprehensively, but the productivity of combining interview, practice-based and stimulated recall methods, particularly with the adaptation of Gendlin's techniques, both supports and responds to calls for the development of diverse methods that allow space for reflecting on practices (Latham 2003), including reflection through re-experience.

Laura talks about having a deep empathy for what she is painting, "a realness, an understanding of what I want to paint", which she describes as an appreciation for "the –ness of everything". Laura says, "I know what the cloudiness of the clouds is to me. There's a certain kind of textural quality to clouds." This is not a specific texture that is expressed only in the texture of her paints, but also in the colours, lustres and forms of the marks made on the canvas (Banfield 2014). While Laura can sum up her empathy for the cloudiness of clouds linguistically, this articulation does not convey the textural qualities that Laura can instil in her artistic work. Reinforcing Susan's earlier comments that she was able to express artistically that which she had not been able to express verbally (but was subsequently able to express that implicit understanding verbally), this suggests again that there are greater limits to the linguistic explication of implicit understanding than to the artistic explication, and that the notion of cloudiness can be used as a conceptual stimulus for an artistically supported process of thinking-at-the-edge in order to explicate cloudiness more fully linguistically.

Artistic practice, through gesture, seemingly provides other-than-linguistic means of conveying implicit understanding, providing one avenue for the potential development of sharp concepts (Gendlin 2006). Complementing this alterity is the potential for artistic practice to act as an intermediate mechanism in bringing implicit understanding to linguistic expression. This was seen most clearly in Susan's linguistic articulation of her experience of serious illness subsequent to artistic articulation, but was also evident in Jane's linguistic improvisation. Lexically speaking, words like scruffle are nonsense words, but for Jane they are full of implicit meaning, brought into artistic and linguistic expression through gesture (Banfield 2014). Linguistically or artistically, the non-representational can be apprehended from two directions. From the first—focusing—we can use our implicit understanding to generate conceptual terms from a stronger implicit base, with artistic practice not only occurring subsequent to focusing, but also functioning as a non-intentional means of focusing, as in Jane's scruffling. From the other direction—thinking-at-the-edge—the representational is always already more-than-representational, such that concepts can be utilized as entry points to the implicit for both artistic and linguistic explication, as in Laura's interrogation of clouds. Artistic practice, it seems, provides its own means of apprehending and conveying the implicit, as well as facilitating the linguistic explication of conceptual understanding from implicit understanding (Banfield 2014). Gendlin's psychotherapeutic methods have the potential to bolster our methodological toolkit with respect to research into affect, and their modified application in this research encourages the combination of real-time practice-based and retrospective recall techniques in such research.

CONSIDERING EXPLICATORY IMPLICATIONS

In this second half of the chapter, I consider seemingly contradictory outcomes from my research, and specifically consider how thinking with Gendlin might help us to account for these apparent contradictions, as well as the implications that such thinking might hold for methodological practice in non-representational geography.

My research generated a complex impression of artistic practices and their resulting spatialities and subjectivities. On the one hand participants expressed surprise at the practices they watched themselves enacting in the video footage, suggesting a certain lack of awareness of their own practices. On the other hand, several participants appeared to possess far greater explicit awareness of their artistic practice while in the throes of doing that practice, compared to their interview accounts. For example, in a previous publication (Banfield 2016a), a consistent pattern was identified in the accounts and practices of three participants, who presented them as strongly rule-governed in their opening interviews, as a mixture of both rule-boundedness and rule-contravention during their practice, and as dismissive of rules and conventions during their review interview. Neither the preliminary nor closing interview accurately conveyed the complexity of their practices as enacted during the production sessions, and their preliminary and closing interview accounts did not concur with one another. In addition, the current research indicated powerful affective experiences of human–nonhuman sociality, which supports findings of previous research that highlighted the attribution by participating artists of control or agency to their materials rather than to themselves (Banfield and Burgess 2013). Such attributions are perceptible, for example, in Jane's comments about the brush enjoying itself, which grant both personification and agency to the brush, and Laura's spatial morphing through her art materials. Such sentiments could perhaps be of relevance to the discrepancy identified between Jane's assertion that she would be able to improvise linguistically while doing her practice and her lack of such linguistic improvisation in the production sessions. I want to consider here how Gendlin's theoretical ideas might help us to account for seemingly very different relative degrees of implicit and explicit understanding of participants' artistic activities during their practice, which complicate common understandings of deep immersion in artistic practice as a loss of self.

Such deep immersion is a common characteristic of a psychological phenomenon called flow (Csikszentmihalyi 1991, 2002). Although its assumption of a psychological subject is not accommodated by non-representational geography, we saw in Chap. 3 how Gendlin's philosophy allows for a human subject that is neither essential nor body-bound. I use flow to open this discussion before complicating conventional understandings of this phenomenon on a Gendlinian basis, through which I also develop a Gendlinian account of the diverse relations between implicit and explicit understanding evident in my research. In brief, flow has been described as the sense of enjoyment gained from the investment of attention beyond the ordinary, which, over time, results in the activity that generated the experience becoming autotelic, or enjoyable for its own sake, independent of any external rewards (Csikszentmihalyi 1975b, 1991, 1996, 2002). In flow, intensive interaction with the environment is said to generate a sense of wholeness when we are totally involved in an activity, in which there is little distinction between self and environment, or past, present and future (Csikszentmihalyi 1975b). Attention is focused on the activity to the exclusion of all else, in a merging of action and awareness as a unified flow of experience (Csikszentmihalyi 1975b, 2002). This lack of distinction between self and environment, and the merging of action and awareness, bring to mind Laura's experience of entering the landscape that she is painting through her materials, suggesting that flow might provide a potential means of understanding such experiences.

There are also points of connection between this understanding of flow and Gendlin's philosophical work, although Gendlin seemingly accounts more fully and persuasively for the diverse degrees of implicit and explicit awareness of artistic practice evident in my research than the idea of flow as deep immersion. As a starting point, the deep absorption in an interaction with the environment and the resulting loss of self-awareness distinct from the environment is reminiscent of Gendlin's body-environment concretion, the primacy of process, and the interaffecting of everything by everything (Gendlin 2001). Flow is said to lead to a more intense interaction with the environment and enhanced sensitivity to the being of others (Csikszentmihalyi and Robinson 1990), which we might associate with "had" space-time. To this extent, perhaps we can consider flow to be a coming to the fore of affective experience, as implicit understanding increasingly takes primacy over explicit understanding. Csikszentmihalyi hints at a similar perspective when describing attention as psychic energy, which determines what does or does not appear to consciousness (Csikszentmihalyi 2002). The focusing of attention through artistic practice might act as a form of focusing in the Gendlinian sense, which can aid access to and apprehension of the implicit, and which is experienced as flow, devoid of self-consciousness. This suggestion will be qualified somewhat later, but, for now, I consider further synergies between Csikszentmihalyi's work on flow and Gendlin's philosophy, particularly in relation to the re-emergent sense of self following a flow experience and the idea of coordinated differentiation from an implicit many.

Although any self-construct is said to become lost in the experience of flow, following a flow experience it is said to be more complex as a result of two processes: first, differentiation, through which the practitioner becomes increasingly unique, for example by the development of enhanced skills; and then, integration, through which the practitioner becomes increasingly unified with other ideas and beings (Csikszentmihalyi 1991, 2002). The combination of these two processes gives a soothing sense of being part of something larger than ourselves, as the holistic sensation of total involvement (Csikszentmihalyi 1975a, b). These paired processes of integration and differentiation are consistent with Gendlin's idea of the coordinated differentiation of entities and objects from the implicit many, while the increasing sense of unification through a flow experience lends Csikszentmihalyi's writings something of a non-representational flavour. The holistic sense of total involvement is reminiscent of Gendlin's interaffecting, of which we become more aware in flow, allowing for a broader understanding of the social to include other ideas, beings, things and materials beyond a humanistic emphasis on a self-contained identity.

The generation of concepts from our implicit understanding also finds equivalent expression in Csikszentmihalyi's accounts of flow. During flow, Csikszentmihalyi proposes that we experience both our uniqueness and our relationship to the cosmos, consequent to which it is easier to generate original thoughts and actions (Csikszentmihalyi 1996).

Notably, the aesthetic flow experience generated in both the doing of art practice and the viewing of art products or visual imagery, has been described as providing four key functions: accessing understanding; generating sensory pleasure; transcending reality; and communicating beyond concepts (Csikszentmihalyi and Robinson 1990). We might take issue with the distinction between reality and what might transcend it, and seek to refine ideas of understanding in terms of implicit and explicit dimensions in light of current Gendlinian interests. However, such propositions are consistent with Gendlin's ideas of communicating from the progressions between concepts, as well as from the concepts themselves, and with the proposal that explicating the implicit can be more effective if an image is allowed to form before words are sought to describe it (Gendlin 1980, 1989, 2009b).

Reading works on flow through Gendlin's core ideas provides a non-representational understanding of the phenomenon of flow in artistic practice and an entry point into an understanding of the variegated artistic practices encountered here, as there is one particular aspect of Csikszentmihalyi's work on flow which, on the basis of the current research, encourages a Gendlinian re-working. Csikszentmihalyi emphasizes the primacy of doing over viewing, proposing that vicarious participation is a pale substitute for the real challenges of practical engagement, suggesting that the focusing of attention (in the sense of flow) might function more effectively as a form of focusing (in the Gendlinian sense) during the practice of creating art than during the viewing of artworks or other images. However, this sits uneasily with the absence of linguistic improvisation in Jane's production sessions, despite her belief that it would be easier for her to do so during her practice. Integrating the ideas of Gendlin and Csikszentmihalyi provides for a more robust account of such findings.

Three core attentional dimensions of flow have been identified which have a bearing on this discussion: the self-construct is lost; attention is object-directed; and the field of stimulus is reduced (Csikszentmihalyi and Robinson 1990). The loss of self-construct in flow has already been discussed in relation to the reported lack of explicit awareness during artistic practice. The object-directedness of attention in flow, in combination with this loss of self-construct, might help us to account for reports of artists attributing agency or control to their materials and equipment as their attention switches from themselves as individual practitioners to the body-environment concretion of their practice (see previous section and Banfield and Burgess 2013). However, the reduction of the field of stimulus said to characterize flow requires rethinking if we seek to bring together Csikszentmihalyi's work on flow with Gendlin's philosophy, as accessing the implicit in flow is perhaps better considered a change in the field of stimulus—from explicit to implicit—rather than a reduction in the stimulus field. With heightened sensitivity to the interaffecting of everything with everything, the field of stimulus could be considered to expand rather than diminish in flow experiences, as our sensitivity to implicit understanding is seemingly elevated above our sensitivity to explicit understanding.

However (and as identified earlier), not all participants exhibited a loss of explicit awareness of their artistic practice during practice, as some appeared to demonstrate greater explicit awareness of their practices while in the throes of doing them than in interview scenarios. One possible reason for this is simply that these participants did not experience flow and its associated loss of self-construct during these production sessions. Conversely, this does not explain why participants' explicit awareness of their practices seemed greater during that practice than outside it (Banfield 2016a). Another possible explanation draws on the idea that artistic practice can function as a form of focusing, increasing participants' access to their implicit understanding during their artistic practice without necessarily losing self-awareness, even if they simultaneously experience flow. This suggests either that the loss of self-awareness in flow is not inevitable or that it does not necessarily lead to a lack of capacity to communicate implicit understanding. With focusing, in contrast to flow, the emphasis is on the rise of implicit awareness, but not to the total exclusion of explicit awareness. If artistic practice functions as a form of focusing, then perhaps the connection between implicit and explicit understanding—so crucial for the explication of the implicit—enables participants to draw on both implicit and explicit understanding during their practice, irrespective of any loss of self-awareness consequent to a simultaneous flow experience. The potential association between focusing and flow in artistic practice, although uncertain, and its implications for geographical concerns with accessing and apprehending the implicit is an intriguing prospect.

The final, seemingly contradictory, set of findings relates to the relative accessibility of implicit understanding between real-time practice and visual recall of practice, as illustrated by Jane's failure to generate linguistic novelties during practice, in contrast to her linguistic improvisation while watching her practice on screen retrospectively. The critical question here is that if, as Csikszentmihalyi and Robinson (1990) suggest, the doing rather than the viewing of art affords greater access to and articulation from the implicit, then why does this accessing of understanding and communicating beyond concepts seem to occur when viewing rather than doing artistic practice?

From a Gendlinian perspective, although the implicit might be accessible and apprehensible during practice, the full power of that experience might not be discernible if the practitioner's self-awareness is lost in flow. Perhaps the full affective power is not discernible until the individual has again become coordinatedly differentiated through the unfolding practice. As a result, the experience might be implicitly felt but not explicitly discerned during flow, only attaining capacity for recognition and communication once a degree of detachment from the implicit many has been re-established, at which point flow would no longer be experienced.

While the capacity to access the implicit during artistic practice is suggested in the accounts and practices of several participants, the degree to which this implicit understanding can be apprehended and communicated seemingly depends upon a specific calibration between implicit and explicit awareness, which emphasizes Gendlin's insistence on a connection between the implicit and the explicit. This might be attainable during practice if explicit understanding can be sustained, irrespective of any loss of self-construct associated with a flow experience. Artistic practice might afford access to implicit understanding, making it available for explication, and which might simultaneously establish the conditions necessary for flow, which might or might not lead to a loss of self-consciousness (with implications for whether the implicit is more palpably sensed and more readily explicated during practice or retrospectively). This appears to suggest a more mediated relationship between artistic practice as focusing and the experience of flow than outlined earlier, which provides fertile territory for further exploration.

On a Gendlinian reading, Jane's commitment to her capacity for linguistic improvisation during her artistic practice is suggestive of practice-based access to her implicit understanding, reminiscent of the proposition that artistic practice can function as a form of focusing. Meanwhile, her lack of linguistic improvisation during the practice-based research sessions— if symptomatic of her practice as a whole—might also suggest a loss of self-consciousness during her practice (consistent with flow), as attention focuses on implicit over explicit understanding. In such a situation, while Jane might have access to her implicit understanding, her lack of explicit self-awareness might preclude any explication of that implicit understanding. However, to generate linguistic novelties from her implicit understanding while reviewing her practice on video, some degree of connection must presumably be re-established with her implicit understanding of her real-time practice, while also retaining her explicit understanding.

Social scientific engagement with video as an affective medium proposes that the telepresence and multi-sensoriality of filmic media allowed the felt sense of her practice to be re-experienced (Cranny-Francis 2009; Featherstone 2010; Garrett 2010; Merchant 2011; Pink 2011a, b; Blackman 2012; Knoblauch 2012; Jacobs 2013, 2015; McHugh 2015). Much of this work draws on psychological work on mirror neurons to account for the affective power of video. Mirror neurons are activated during a motor action (e.g. raising an arm) and are also stimulated when that action is observed (Calvo-Merino et al. 2005; Aziz-Zadeh and Ivry 2009; Iacoboni 2009; Semin and Cacioppo 2009). These neurons are proposed to provide access to motor and associated affective registers and responses through a synaesthetic–kinaesthetic convergence, which enables the re-experiencing of that which had been formerly present and the inter-subjective conveyance of affect (Gibbs 2009; Iacoboni 2009; Rochat and Passos-Ferreira 2009; Winkielman et al. 2009; Blackman 2010, 2012; Featherstone 2010). However, it remains unclear why something that can be articulated while watching one's own practice was incapable of articulation during the practice that is being watched. Gendlin also helps to inform our understanding of this apparent phenomenon, and again, Jane's scruffling is exemplary here. If Jane re-experienced her original practice during retrospective recall, why could she articulate something during recall that she had not articulated during practice?

Earlier, I emphasized Gendlin's work on the connections between implicit and explicit understanding and the possibility that for the implicit to be explicated during practice some degree of self-awareness might be necessary, such that, if self-awareness was lost in flow, such explication during practice would be precluded. When watching one's own actions on film, the video might function to sustain a sufficient degree of self-consciousness to facilitate the explication of implicit understanding, while simultaneously empathetically or mimetically re-establishing the experience of practice previously experienced. While the direct previous personal experience of that practice—especially if under conditions of flow—perhaps optimizes the extent to which implicit understanding is available for explication, deep immersion in that practice (as in flow) perhaps fails to sustain sufficient explicit awareness to support explication. The visual image of one's self on video, while affording a multi-sensory re-experience of previous practice, might also sustain the practitioner's awareness of themselves as coordinately differentiated from the implicit many, within which they might have been too embedded during real-time practice to enable explication of implicit understanding at that time.

It is therefore feasible that implicit understanding might be more readily explicated in retrospective recall (where explicit understanding is sustained more effectively than during practice), if during that practice implicit understanding is elevated above explicit understanding to a sufficient extent. Consequently, in accordance with Gendlin's emphasis on the connection between the implicit and the explicit, the key to accessing, and particularly apprehending, the implicit is not so much focusing on the implicit over the explicit (as might be the case during practice) but on enhancing sensitivity to the implicit while retaining access to the explicit (as might be the case with video recall). Such a possibility perhaps supports research designs that employ both real-time practice-based methods and visually stimulated recall methods as potentially effective methods for accessing and apprehending the implicit, which might be further enhanced by the application of Gendlin-inspired explicatory techniques.

CONCLUSION

This chapter has detailed my own methodological application of adapted versions of Gendlin's explicatory techniques and the outcomes of those efforts. With Laura, these methods generated more detailed accounts of the experiential and affective aspects of her artistic practice, raised questions with regard to the possible impact of handedness on explication, and pointed to the potential for using art-based approaches to thinking-at-the-edge to refine implicitly linguistic descriptors. With Jane, these methods stimulated linguistic improvisation, perhaps indicating that she was speaking directly from her implicit understanding, and raised questions concerning the relation between doing, viewing and saying. This confused relation is symptomatic of a number of potentially contradictory outcomes of my research, which revolve around the relative accessibility of implicit and explicit understanding in different circumstances. Gendlin afforded a way to account for these diverse findings, while also encouraging us to reconceptualize psychological understandings of flow.

Flow was associated with an increased sensitivity to the interaffecting of everything by everything, but was problematized as potentially precluding the articulation of implicit understanding during practice. Optimal explication was suggested as arising from the simultaneous maintenance of explicit awareness and enhancement of implicit awareness, allowing for the intentional exploitation of their connectivity. This suggests a particular interaction between flow and focusing in artistic practice, and potentially accounts for the different degrees of implicit/explicit awareness evidenced in different circumstances of practice.

We can also (in a Gendlinian sense) conceptualize the real-time practice-based aspect of my methodology in terms of focusing (increasing access to the implicit) and the retrospective stimulated recall methods in terms of thinking-at-the-edge (connecting implicit and explicit understandings). It is this state of connectivity between the implicit and explicit during embodied simulation of affective experience in video-elicitation that might provide optimal conditions for the explication of sharp concepts from implicit understanding, allowing the re-experience of a former felt sense, while simultaneously sustaining sufficient explicit understanding, and potentially making these combined methods a powerful means of researching affect.

However, these varied research findings suggest interpersonal differences in the relative accessibility of the implicit, and ease and effectiveness of explication, and also encourage the application of a range of research methods tailored for specific participants. Geographical research into affect might benefit from adopting longer term and multi-stage research designs, in which initial work identifies the most effective method/s for each participant in advance of the substance of the research being undertaken, emphasizing focusing through practice for some, thinking-at-the-edge through video-elicitation for others, and supplementing either or both with Gendlinian explicatory techniques as appropriate. This is an issue to which I return in the next chapter, which presents a critical review of my engagement with Gendlin.

REFERENCES

Aziz-Zadeh, Lisa & Ivry, Richard B 2009 The human mirror neuron system and embodied representations. *in:* Sternad, D (ed.) *Progress in motor control.* Springer, New York; London 355–376

Banfield, Janet 2014 *Towards a non-representational geography of artistic practice.* Unpublished doctoral thesis, University of Oxford, Forthcoming online: https://ora.ox.ac.uk:443/objects/uuid:dd12e1c4-f222-435b-adc0-c1bb68e4f4ac

Banfield, Janet 2016a Knowing between: generating boundary understanding through discordant situations in geographic-artistic research. *Cultural Geographies*, 23 459–473

Banfield, Janet & Burgess, Mark 2013 A phenomenology of artistic doing: flow as embodied knowing in 2D and 3D professional artists. *Journal of Phenomenological Psychology* 44 60–91

Blackman, Lisa 2010 Embodying affect: Voice-hearing, telepathy, suggestion and modelling the non-conscious. *Body and Society* 16 163–192

Blackman, Lisa 2012 *Immaterial bodies: affect, embodiment, mediation.* SAGE, London

Calvo-Merino, B, Glaser, DE, Grezes, J, Passingham, RE & Haggard, P 2005 Action observation and acquired motor skills: an fMRI study with expert dancers. *Cerebral Cortex* 15 1243–1249

Cranny-Francis, Anne 2009 Touching film: the embodied practice and politics of film viewing and filmmaking. *Senses and Society* 4 163–178

Csikszentmihalyi, Mihalyi 1975a *Beyond boredom and anxiety: (the experience of play in work and games).* Jossey-Bass, San Francisco; London

Csikszentmihalyi, Mihalyi 1975b Play and intrinsic rewards. *Journal of Humanistic Psychology* 15 41–63

Csikszentmihalyi, Mihalyi 1991 *Flow: the psychology of optimal experience.* Harper Perennial, New York; London

Csikszentmihalyi, Mihalyi 1996 *Creativity: flow and the psychology of discovery and invention.* Harper Collins, New York

Csikszentmihalyi, Mihalyi 2002 *Flow: the classic work on how to achieve happiness.* Rider, London

Csikszentmihalyi, Mihalyi & Robinson, Rick E 1990 *The art of seeing: an interpretation of the aesthetic encounter* J.P. Getty Museum, Getty Center for Education in the Arts, Los Angeles

Drew, Nancy 2006 Bridging the distance between the objectivism of research and the subjectivity of the researcher. *Advances in Nursing Science* 29 181–191

Featherstone, Mike 2010 Body, image and affect in consumer culture. *Body and Society* 16 193–221

Garrett, Bradley L 2010 Videographic geographies: Using digital video for geographic research. *Progress in Human Geography* 35 521–541

Gendlin, Eugene T 1980 Imagery is more powerful with focusing: theory and practice. *in:* Shorr, JE, Sobel, GE, Robin, P & Connella, JA (eds.) *Imagery. Its many dimensions and applications.* Plenum Press, New York; London 65–73

Gendlin, Eugene T 1989 Phenomenology as non-logical steps. *in:* Kaelin, EF & Schrag, CO (eds.) *American Phenomenology: origins and developments.* Kluwer, Dordrecht 404–410

Gendlin, Eugene T 2001 *A Process Model* The Focusing Institute, New York

Gendlin, Eugene T 2006 Transcript of Gendlin Templeton Lecture. Psychology of Trust and Feeling Conference. Stony Brook University. http://www.focusing. org/gendlin_templeton.html. Accessed 05 Nov 2012

Gendlin, Eugene 2009a What first and third person processes really are. *Journal of Consciousness Studies* 16 332–362

Gendlin, Eugene T 2009b We can think with the implicit, as well as with fully-formed concepts. *in:* Leidlmair, K (ed.) *After cognitivism: a reassessment of cognitive science and philosophy.* Springer, London, New York 147–161

Gibbs, Anna 2009 After affect: sympathy, synchrony and mimetic communication. *in:* Gregg, M & Seigworth, GJ (eds.) *The affect theory reader.* Duke University Press, Durham, NC; London 186–205

Hitchings, Russell 2012 People can talk about their practices. *Area* 44 61–67

Iacoboni, Marco 2009 The problem of other minds is not a problem: mirror neurons and intersubjectivity. *in:* Pineda, JA (ed.) *Mirror neuron systems.* Humana Press, New York 121–133

Ingold, Tim 2007 *Lines: a brief history* Routledge, London

Jacobs, Jessica 2013 Listen with your eyes; towards a filmic geography. *Geography Compass* 7 714–728

Jacobs, Jessica 2015 Visualising the visceral: Using film to research the ineffable. *Area* doi: 10.1111/area.12198

Knoblauch, Hubert 2012 Videography. *in:* Knoblauch, H, Schnetter, B & Raab, J (eds.) *Video analysis: methodology and methods: qualitative audiovisual data analysis in sociology,* 3rd edition. Lang, Frankfurt; Oxford 69–83

Latham, Alan 2003 Research, performance, and doing human geography: Some reflections on the diary-photograph, diary-interview method. *Environment and Planning A* 35 1993–2017

Manning, Erin 2009 *Relationscapes: movement, art, philosophy* MIT, Cambridge, MA; London

McHugh, Kevin E 2015 Touch at a distance: toward a phenomenology of film. *GeoJournal* 80(6), 839–851.

Merchant, Stephanie 2011 The body and the senses: Visual methods, videography and the submarine sensorium. *Body and Society* 17 53–72

Pink, Sarah 2011a Images, senses and applications: Engaging visual anthropology. *Visual Anthropology* 24 437–454

Pink, Sarah 2011b A multisensory approach to visual methods. *in:* Margolis, E & Pauwels, L (eds.) *The SAGE handbook visual research methods.* SAGE, London 601–615

Rochat, Philippe & Passos-Ferreira, Claudia 2009 From imitation to reciprocation and mutual recognition. *in:* Pineda, JA (ed.) *Mirror neuron systems.* Humana Press, New York 191–212

Semin, Gün R & Cacioppo, John T 2009 From embodied representation to co-regulation. *in:* Pineda, JA (ed.) *Mirror neuron systems.* Humana Press, New York 107–119

Stelter, Reinhard 2010 Experience-based, body-anchored qualitative research interviewing. *Qualitative Health Research* 20 859–867

Tallis, Raymond 2003 *The hand: a philosophical enquiry into human being* Edinburgh University Press, Edinburgh

Tallis, Raymond 2004 *I AM: a philosophical inquiry into first-person being* Edinburgh University Press, Edinburgh

The Focusing Institute, www.focusing.org

Winkielman, Piotr; Niedenthal, Paula M & Oberman, Lindsay M 2009 Embodied perspective on emotion-cognition interactions. *in:* Pineda, J A (ed.) *Mirror neuron systems.* Humana Press, New York 235–257

Critiquing Explicatory Techniques

Abstract Chapter 7 contains a self-critical evaluation of the application of methods informed by Eugene Gendlin's techniques within geographical research. Banfield identifies and discusses potential shortcomings in methodological choices and characterizes the research design itself in Gendlinian terms. On this basis, she constructively reframes two potentially conflicting aspects of the research design as complementary rather than contradictory, encouraging further investigation of and through such research designs. She also explores the implications for ongoing methodological innovation, both within and beyond non-representational geography, arising from this discussion and reinforces calls for individually tailored research designs. Through this critical evaluation, Banfield underlines the potential contribution that Gendlin's work can make to geography.

INTRODUCTION

In this chapter, I reflect more critically on my experimental adaptation and application of Gendlin's explicatory methods. I attend particularly to the demographic and practice characteristics of the participants, the manner in which I adapted Gendlin's specific methods and the broader research setting within which these methods were devised. I consider the implications of my engagement with Gendlin's work, in relation to what Gendlin's

© The Author(s) 2016
J. Banfield, *Geography Meets Gendlin*,
DOI 10.1057/978-1-137-60440-8_7

methods offer over and above existing artistic and video-based research methods and, more specifically, with regard to existing research practice employing video-based recall methods. I argue that despite difficulties discovered in relation to the research design—which might have inadvertently and detrimentally affected the effectiveness of these Gendlin-informed methods—the research findings hint at a greater potential for such methods to contribute to non-representational geography than is immediately apparent. I also outline changes to the research design which might facilitate the actualization of this potential contribution, inviting and stimulating further work in this area.

Participant Characteristics

Given the small and selective sample of participants with whom I worked, this research is neither conclusive nor representative (Banfield 2014). On the one hand, all participants are female, all except one were over the age of 50, and all of those who practised professionally had taken up their artistic practice as a second career. None of these characteristics had been introduced intentionally into the research design, but while there is no evidence of any particular impact that these features had on the research, the findings might have been very different with a different group of participants. On the other hand, although all except two of the participants practised their art professionally, the length and status of art career, the nature, level and location of formal training, the art traditions, media and styles that participants practised all varied across the sample. Such diversity in the nature of participant practices makes it difficult to isolate and identify any particular influences on the research findings that might result from the types of materials used, the art traditions followed, or the duration of practice dispositions. In addition, irrespective of these practice variations, all participants, by virtue of their involvement in arts-related activities, function within a certain community of practice, which in turn requires a certain artistic and social literacy and confidence (Banfield 2014). As a result, I am unable to contribute to debates as to whether such explicatory techniques, combined with real-time practice-based research, would generate similar findings in other socio-cultural contexts, in other types of practice, or through artistic practices undertaken by people who do not voluntarily engage with artistic practice. Even so, as my intention was to explore particular features of varied artistic practices in specific circumstances, rather than to generate a generalizable account, theory or model of those practices, such specificity is not particularly problematic. My concern is with Gendlin's potential contribution, rather than with the substantive research findings. Given these considerations, I offer indications of the potential for Gendlin's philosophy and techniques to inform geographical research based upon my experimental engagement with his work. I also suggest possible avenues for further exploration of this potential in non-representational geographies, geographies of artistic practice and more-than-human geographies.

Demographic differences alone might be sufficient to account for the variability evident in participants' access to and explication of implicit understanding, as well as differences in sensitivity to pre-reflective experience, whether due to variations in somatic knowledge, cultural differences, or differences in the circumstances of encounter and display (Rose 2003; Howes 2006; Cranny-Francis 2009). However, these individual differences are no doubt compounded by practice differences among participants. At this point I return to the concept and function of mirror neurons, not to assume unquestioningly that mirror neuron function is the mechanism by which intersubjective affective understanding is attained, but to complicate significantly any such straightforward assumptions. Rather than risk a mistranslation of psychological knowledge into social science practice and understanding (Clough 2010), I seek instead to stimulate further critical interdisciplinary cross-fertilization.

In particular, the proposed mirror neuron function has potential implications with regard to the level of training, proficiency or expertise of research participants, as the translation of visual input into motor activation patterns has been reported to occur in accordance with the acquired skill of the observer. In other words, the brain's response to seeing an action performed by another person is influenced by the acquired motor skills of the observer; activation during motor observation is stronger in dancers for actions that were within their own repertoire (Calvo-Merino et al. 2005). This suggests perhaps that accessing and apprehending pre-reflective aspects of experience intersubjectively might benefit from engaging expert rather than beginner or learner practitioners, as they would be responsive, and more so, to a greater range of pre-reflective experiences. However, supporting Harriet Hawkins' advocacy of avoiding fetishizing skill or confining such practice-based research to experts (Hawkins 2015), the influence of acquired motor skills on mirror neuron function also brings other possibilities.

For example, it suggests that accessing and apprehending pre-reflective aspects of experience on an individual basis might be particularly effective when looking at one's own recorded activity, irrespective of proficiency, as this offers the greatest likelihood that the actions observed are within the observer's repertoire. It also indicates interesting avenues of inquiry on an intersubjective basis between practitioners of equivalent proficiency, irrespective of their formal level of expertise. Similarly, while the reported role of acquired skill in mirror neuron function might encourage research in which the observer practices the same activity that is being observed, research conducted on a comparative basis between different practices might yield interesting insights into the embodied experiences of those practices and their intersubjective intelligibility, as was suggested in relation to the intelligibility of Jane's scruffling as an avenue for further investigation.

Another specific issue related to the potential role of mirror neurons concerns lateralization, or preferential activation, of neural circuits on one or other hemisphere of the brain in the execution of certain tasks or in response to certain stimuli. For example, the suggestion that the left hemisphere might encode more abstractly or multi-sensorially than the right hemisphere, which is more modality specific and limited (Aziz-Zadeh and Ivry 2009), is of particular interest in light of Laura's ambidextrous artistic practice and her different articulations of the actions of each hand during visually stimulated recall. Issues such as left or right dominance in handedness are likely to interact in highly complex ways with individual differences in sensitivity, practice and proficiency. Interestingly, mirror neurons have also been suggested as potentially holding a key to understanding the neural basis of conceptual knowledge through the interpretation of the actions of others and inferring others' intentions (Aziz-Zadeh and Ivry 2009; Iacoboni 2009). Reminiscent of Gendlin's implicit–explicit connectivity, such embodied cognition theories assume that high level cognitive processes involve sensory re-activations (Winkielman et al. 2009). These potential connections between an emerging understanding of mirror neurons and Gendlin's philosophy encourage more critical social science engagement with psychological concepts, such as mirror neurons. It also reinforces the advocacy in the previous chapter of more individually tailored research designs, which employ varied methods to cater for participant diversity. Finally, it suggests particular benefits of combining Gendlin-informed and video-elicitation techniques in the explication of implicit understanding.

METHODOLOGICAL ADAPTATIONS

The most significant shortcomings of my exploration of disciplinary potential within Gendlin's methods relate to research design. I adopted two primary approaches to attend to the pre-reflective aspects of participants' artistic practices. In addition to observing participants' practices as they would normally undertake them, the first approach involved establishing research scenarios in which participants were asked to change material and/or spatial aspects of their practices. The aim of this approach was to draw participants' attention to aspects of their practice that they might normally overlook through a comparison between their normal and changed practices (for details of this approach, see Chap. 2 and Banfield 2014, 2016a). Through comparison, I sought to support participants in articulating from their implicit understanding during the production sessions, in what could be considered a practical equivalent to Gendlin's crossing of concepts. The second approach involved the experimental application of Gendlin-informed explicatory techniques in the closing interview, as detailed in the previous chapter. This two-pronged methodology introduced at least two significant problems. The first and most straightforward issue relates to my adaptation of Gendlin's explicatory techniques; the second and more substantive issue is a possible inherent conflict between the two prongs of my methodology. Both demand critical consideration.

I did not employ Gendlin's techniques as Gendlin originally specified, for two primary reasons. First, I was concerned that the length of training that is deemed necessary for the effective use of explicatory techniques in therapeutic situations could discourage participation. Second, the focus of Gendlin's training system on the verbal explication of implicit understanding contrasted with my focus, in the first instance at least, on explication through artistic practice. In the absence of formal therapeutic aims that might demand such lengthy training, and with a focus on potential for artistic rather than linguistic explication within practice, I was keen to streamline Gendlin's techniques as far as possible, while still providing a means of testing whether they hold promise for non-representational geography (Banfield 2014). In addition, I drew on alternative interview techniques, including re-enactment and visually stimulated recall, to provide a number of potential routes into the implicit. Such modifications complicate the attribution of research outcomes specifically to Gendlin's core techniques. With hindsight, and given the variability in research findings, a greater degree of researcher and participant training might have proved more fruitful. However, the elicitation of more detailed accounts of artistic practices through Gendlin-informed interview techniques of contracting and expanding the intended meaning, and the generation of linguistic improvisation during visually stimulated recall, in the absence of any formal training suggests that such training is not always necessary. At the very least, the use of Gendlinian techniques in a streamlined fashion can prevent reported difficulties in narrative accounts of visually recalled practices being broad and superficial (Merchant 2011), by directing participant engagement with visually stimulated recall material. Similarly, Jane's linguistic improvisation during visually stimulated recall suggests that such recall methods can be an effective supplementary tool for interviews aimed at explicating implicit understanding. Streamlined versions of Gendlin's techniques, then, have the potential to generate fuller interview data than might be the case with conventional qualitative interviews, irrespective of any focus on affect, while other recall-based interview techniques have the potential to support Gendlinian techniques aimed specifically at explicating implicit understanding.

The second obstacle introduced by my two-pronged methodology—an inherent conflict between the two prongs—is more serious, and has only come to light in writing this book, which is more tightly focused on the potential contribution of Gendlin to non-representational geography than the original and broader research. Specifically, it is through thinking about the potential function and implications of mirror neurons that this issue has come to light and, put simply, the problem is as follows. If the potential value of mirror neurons to research such as this arises from either (A) expert or proficient practices, or (B) practices within the acquired repertoire of the performer, then any method which changes the observed practices from acquired or accomplished experience to novel or unusual experiment—which is precisely the nature of my first methodological prong—will surely detrimentally influence the effectiveness of visually stimulated recall methods that draw on mirror neuron function.

Within my overall research aims, both methodological prongs were established with the same intention of maximizing opportunities for participants to become aware of aspects of their practices of which they might not normally be aware, but the first potentially undermined the latter. However, other aspects of mirror neuron function serve to mitigate the severity of this potential conflict between methodological approaches. On the one hand, any mirror neuron function and influence that might be dissipated by the non-habitual nature of the skills being observed would be mitigated by the observation of the participant's own practice; even though those practices were atypical, they were still clearly within the repertoire of the participants, as they had performed them. On the other hand, it seems far from clear at this stage how mirror neurons might function when viewing atypical actions within the observer's own practice, raising the possibility that a strong contrast between habitual and unusual practice might trigger mirror neuron activation in such a way that explication from the implicit is facilitated rather than hampered. With the changes introduced to participant practices conceptualized as a practical form of crossing, the two methodological prongs are perhaps not so much conflicting as complementary, with the first approach of varying practice supporting the second approach of experimental explicatory methods. Consequently, an issue which at first seemed potentially catastrophic is instead an intriguing conundrum that invites further investigation.

Taken together, the variability evident in participant characteristics and practices, and the diversity evident in the nature and extent of explication from implicit understanding arising from the modification and combination of the techniques employed, argues strongly in favour of multiple, flexible and tailored methods in research into pre-reflective experience. As evidenced in the comparison drawn between the accounts of Laura and Jane, different participants responded uniquely to similar tasks, and communicated their implicit understanding in very different ways. Laura communicated affective qualities of her practice quite freely during the production sessions, and employed conventional language during the stimulated recall exercise in the closing interview; whereas Jane communicated the affective qualities of her practice in the stimulated recall exercise and spontaneously generated linguistic novelties only during this exercise, despite her belief that this would happen during her real-time practice. If participant sensitivity to the implicit and responsiveness to different explicatory techniques can be discerned, research efficiency and effectiveness could be maximized, while simultaneously minimizing the burden on participants in terms of the time invested in both training on explicatory techniques and the research itself.

GENDLINIAN IMPLICATIONS

Thinking through the research method and findings in the context of Gendlin's writings, certain implications arise with respect to prevailing understandings of video practice, and which also raise additional critical issues with regard to my own research.

The integrated use of real-time artistic practices and the visually stimulated recall of those practices, in combination with Gendlin-informed explicatory techniques, constitute a formative methodological development with potential for further exploration and critique. As outlined earlier, emerging psychological understandings of mirror neurons can inform our video-based research practices, especially as a potential means of inquiry into implicit understanding, intersubjective affectivity and more-than-human sociality.

I also considered, in the last chapter, consistencies between the concept of flow and Gendlin's philosophy, which further elucidated some of Gendlin's core terms and ideas, and which culminated in a suggestion that it might be appropriate to reconsider commonly described features of flow along Gendlinian lines. Finally, I proposed that my own research method can be described in Gendlinian terms, whereby the comparison drawn between customary and atypical practices was characterized in terms of crossing, through which further implicit meaning is made available for explication. The real-time practice elements can be considered an artistic application of focusing, the stimulated recall aspects can be considered a video-based application of thinking-at-the-edge, and their combined application could provide a particularly effective means of researching the implicit by optimizing the simultaneous availability of implicit and explicit understanding.

Earlier, I discussed the caveats around social scientific employment of physical science knowledge (Clough 2010) and the complications injected into this interdisciplinary engagement by this research. However, the suggestion that the activation of mirror neurons enables us to access the more-than-visual brings with it implications for our understanding of the content of video recordings and the most appropriate means of working with or analysing them.

The capacity for the generation of more detail, significance or alternative interpretations in reviewing video footage is well rehearsed, and it is not unusual for authors to advocate concern only for a transparent reading of the manifest (visually apparent) content of a video recording, specifically arguing against speculating with regard to what does or does not appear in the video (see, for example, Knoblauch et al. 2006; Banks 2007; Laurier and Philo 2012). In light of the earlier discussion of video as an affective medium, though, it is uncertain where such edicts leave the more-than-visual that is reportedly experienced during video-elicitation even though it is not evident in the manifest content of the video footage.

A distinction might be useful here between what is identified in the video by participants and what is identified by the researcher, especially where the participants are reviewing film of their own practices, and perhaps even more so if participants are experts in those practices. Non-expert researchers might be less sensitive to the multi-sensorial activation triggered by the video than expert participants, because, unlike expert participants, they would not have the same repertoire of practices and experiences as the person being observed on the video. Consequently, participants might be best placed to identify meaningful footage through which the more-than-visual might be accessed.

The issue of specification as to what is significant in a video record-ing raises another potential limitation of my research, as I conducted the initial analysis of the video recordings to identify and isolate short clips for review in the closing interviews. As my repertoire of artistic practice is markedly different to that of the participants, I might have identified clips for review that participants would not have selected themselves. While the meaningful responses that participants did generate to the selected video clips illustrate that researcher selection of video clips for visually stimulated recall purposes is far from devoid of value, the potential for more effective explication through participant-selected stimulus material needs further investigation.

Another particular concern raised with regard to the use of visually stim-ulated recall of practices as a research method is the risk that the responses generated by viewing a video of previous practice would be framed by the contemporaneous experience of viewing the video, rather than the original experience of the practice that is shown (Banks 2007). However, while mirror neuron function might mitigate such concerns by enabling the simulated re-experience of that action on a pre-reflective level, it is also possible to consider the situated nature of the viewing of the video more affirmatively. Drawing upon Gendlin, the situated circumstances when viewing the video might aid the maintenance of explicit understand-ing, while also re-establishing the original felt sense, creating a particular potential for research into affect. In addition to the internal narrative of the manifest visual content and two external narratives—in relation to the circumstances of the original event and to the circumstances of the viewing of the video (Banks 2007)—the existence of a second internal narrative beyond the manifest visual content is suggested by ideas of video as an affective medium. It is worth considering the potential role and value of the external narrative related to viewing the video as a means of accessing and articulating that second internal narrative, by establishing the appropriate calibration between implicit and explicit understanding to enable explication from the implicit. Consequently, the potential within Gendlin's methods, either as originally specified or as subsequently adapted, in combination with real-time practice-based and retrospective stimulated recall methods, warrants deeper engagement and consideration both within geography and beyond.

CONCLUSION

In concluding this more critical consideration of the explicatory techniques and broader research methods that I employed, it is fair to say that certain features of the research context, such as the demographic and practice features of the participant group, undoubtedly influenced the research findings. However, some of these (e.g. individual differences) are unavoidable in any research with human participants, while others are associated with my research intentions and aims (e.g. its exploratory nature). There are undeniably aspects of my research that I could have executed differently, and which might have had a significant influence on the research data and findings. While some of the choices that I made brought their own benefits (e.g. streamlining Gendlin's methods), some also potentially undermined other aspects of my methodology (e.g. changing the nature of participants' practices). However, it is also possible that these different approaches to accessing aspects of participants' artistic practices of which they would not normally be explicitly aware, functioned as, or more, effectively in combination than if I had used Gendlin-informed explicatory methods alone. It is therefore difficult to say with any certainty whether these particular methodological choices, on balance, helped or hindered the effective application of these methods.

What it is possible to say is that my methodological choices generated a number of interesting and meaningful outcomes. They generated fuller accounts of the artistic practices encountered, specifically concerning the pre-reflective aspects of those practices, which raised questions concerning the influence of factors such as lateralization on explication. They also highlighted linguistic improvisation as a potential route to apprehending and communicating implicit understanding, which identified potential avenues for further investigation. With a clarification of methodological implications, future engagement with Gendlin's methods of focusing and thinking-at-the-edge, whether as originally specified or as adapted, might yield considerable benefit for non-representational geography, geographical research employing video methods, geographies of artistic practice and, to a lesser degree, more-than-human geographies. It is also possible to say that these findings allow both a further elucidation of Gendlin's core ideas in the context of geographical interest in artistic practice, and a reconceptualization of both geographic and artistic practices in Gendlinian terms. Beyond geography, these methodological choices and the theoretical discussion surrounding the research findings stimulate a rethinking of geographical engagement with psychological phenomena (flow and mirror neurons) through Gendlin's ideas and concepts. As a result, a Gendlin-inflected geography holds considerable promise for contributing significantly to conceptual understanding and methodological practice far beyond geography itself.

References

Aziz-Zadeh, Lisa & Ivry, Richard B 2009 The human mirror neuron system and embodied representations. *in:* Sternad, D (ed.) *Progress in motor control.* Springer, New York; London 355–376

Banfield, Janet 2014 *Towards a non-representational geography of artistic practice.* Unpublished doctoral thesis, University of Oxford, Forthcoming online: https://ora.ox.ac.uk:443/objects/uuid:dd12e1c4-f222-435b-adc0-c1bb68e4f4ac

Banfield, Janet 2016a Knowing between: generating boundary understanding through discordant situations in geographic-artistic research. *Cultural Geographies,* 23 459–473

Banks, Marcus 2007 *Using visual data in qualitative research.* SAGE, Los Angeles; London

Calvo-Merino, B, Glaser, DE, Grezes, J, Passingham, RE & Haggard, P 2005 Action observation and acquired motor skills: an fMRI study with expert dancers. *Cerebral Cortex* 15 1243–1249

Clough, Patricia T 2010 Afterword: The future of affect studies. *Body and Society* 16 222–230

Cranny-Francis, Anne 2009 Touching film: the embodied practice and politics of film viewing and filmmaking. *Senses and Society* 4 163–178

Hawkins, Harriet 2015 Creative geographic methods: knowing, representing, intervening. On composing place and page. *Cultural Geographies* 22 247–268

Howes, David 2006 Charting the sensorial revolution. *Senses and Society* 1 113–128

Iacoboni, Marco 2009 The problem of other minds is not a problem: mirror neurons and intersubjectivity. *in:* Pineda, JA (ed.) *Mirror neuron systems.* Humana Press, New York 121–133

Knoblauch, Hubert; Schnetter, Bernt & Raab, Jürgen 2006 *Video analysis: methodology and methods: qualitative audiovisual data analysis in sociology*, 3rd edition. Lang, Frankfurt; Oxford

Laurier, Eric & Philo, Chris 2012 Natural problems of naturalistic video data. *in:* Knoblauch, H, Schnetter, B & Raab, J (eds.) *Video analysis: methodology and methods: qualitative audiovisual data analysis in sociology*, 3rd edition. Lang, Frankfurt; Oxford 181–190

Merchant, Stephanie 2011 The body and the senses: Visual methods, videography and the submarine sensorium. *Body and Society* 17 53–72

Rose, Gillian 2003 On the need to ask how, exactly, is geography "visual"? *Antipode* 35 212–221

Winkielman, Piotr; Niedenthal, Paula M & Oberman, Lindsay M 2009 Embodied perspective on emotion-cognition interactions. *in:* Pineda, J A (ed.) *Mirror neuron systems.* Humana Press, New York 235–257

CHAPTER 8

Conclusion

Abstract The conclusion details key contributions that Eugene Gendlin's work can make to geography and beyond, as identified through geographical research into artistic practices. Banfield highlights anticipated contributions to philosophical and methodological debates within non-representational geography, and the emergence of a new terminology and new ways of thinking about geographies of artistic practice. She proposes that this geographical engagement with Gendlin develops recent disciplinary efforts to rethink abstraction affirmatively, and has the potential to enhance the disciplinary capacity to work conceptually with images. She argues that it also: informs evolving disciplinary thinking and practice involving video-based methods; encourages individually tailored research designs to access the affective; and identifies potential means by which we can develop our capacity for working with the pre-reflective intersubjectively.

Introduction

The aim of this book was to introduce the philosophical and psychotherapeutic work of Eugene Gendlin to geography, in order to invigorate debate within non-representational geography concerning human subjectivity and inform methodological innovation in its efforts to access and apprehend the implicit (affect). I intended to flesh out Gendlin's key terms and

© The Author(s) 2016
J. Banfield, *Geography Meets Gendlin*,
DOI 10.1057/978-1-137-60440-8_8

ideas within the context of my own geographical research into the spatialities and subjectivities that emerge through artistic practice. I drew on the
accounts and practices of participating artists, supplemented by my own
hobby artistic practice, to illustrate and explore key ideas from Gendlin's
philosophy, and both documented and critiqued my own experimental
application of research methods derived from Gendlin's psychotherapeutic techniques. The challenge for this final discussion is to condense the
outcomes and implications of the foregoing chapters into an account of
the key contributions that Gendlin might make to non-representational
geography and to identify further potential in broader areas of academic
inquiry, both geographical and otherwise.

CONTRIBUTIONS

Part 1 identified the interdisciplinary interests, themes and challenges
that informed my own research, through which I explored Gendlin's
philosophy and methods. In Chap. 1, I focused on non-representational
geography's emphasis on affect, and its specific concerns with notions
of human subjectivity and how we might access and apprehend affect. I
highlighted connections and distinctions between Gendlin's philosophy
and current understandings within non-representational geography, paying particular attention to the relation between the representational and
the non-representational, to pave the way for geographical engagement
with Gendlin's work. In Chap. 2, I discussed the increasing convergence
of research approaches and methods across the social sciences, illustrating
these developments in the context of geographical engagement with artistic products and practices. I highlighted the coming together of artistic,
practice-based, psychological and video-elicitation methods, which offers
potential for the development of methods to access embodied experience and knowledge. I also described how these developments informed
my own research design, which combined these methods with Gendlin-
informed interview techniques. Chapter 2 brought together themes and
methodological challenges from non-representational geography and the
geographies of artistic practice to which my research responded, setting
the scene for Parts 2 and 3, which detailed my exploration and evaluation
of Gendlin's potential contribution to geography.

Each chapter in Part 2 adopted a key idea from Gendlin's philosophy as
an orienting device. Chapter 3 took implying and its relation to occurring
as a point of entry to the exploration of his conceptual material through

empirical data from my research, and brought together Gendlin's ideas about temporality and "had" space-time to work through the potential for non-representational geography to accommodate a human subject. This discussion provides a new philosophy and terminology through which to think non-representationally, and a new way to think about geographies of artistic practice, highlighting the emergence of spatiality and subjectivity through an indeterminate sequence of implyings and occurrings. The potential for Gendlin's philosophy to provide for a non-representational subject has particular value for geography, which is currently grappling with this very question. More broadly, my exploration of Gendlin's thinking in relation to coordinated differentiation, objects and processes has the potential to inform more-than-human geographical interest in human–nonhuman relations.

Chapters 4 and 5 both attended to the relation between the explicit (representational) and the implicit (non-representational), and contribute to emerging concerns within geographies of artistic practice for the discipline's capacity to think about images and image-making in a conceptual manner (Hawkins 2015), but each did so through a different focus. Chapter 4 adopted the idea of explication—the development or lifting out of explicit or conceptual understanding from pre-reflective or implicit understanding—to address another contemporary concern for non-representational geography, whether we are able to apprehend affect in representational form without losing its affective power. This chapter explored the potential for the explication of implicit understanding in both verbal and visual forms, paying particular attention to the suggestion that visual explication can aid subsequent verbal explication. However, through a detailed consideration of the integration of symbolism and narrative in the practices of participating artists, I also raised questions concerning the intersubjective intelligibility of certain artistic explications. For such explications, subsequent linguistic explication might be required if artistic explicatory methods are applied in research scenarios, possibly limiting their applicability in the field. Subsequently, I highlighted the potential within Gendlin's idea of explication for enhancing access to and articulation from implicit understanding, and some potentially significant limitations in the context of geographical field research, which invite further investigation.

Chapter 5 focused on progression as the implicit connectivity between supposedly distinct concepts. Progression here was characterized as a latent interaffecting, capable of being activated by Gendlinian practices of focusing, dipping, crossing and thinking-at-the-edge. These practices contribute to the explication of implicit understanding, which we can understand as an intentional amplification of the progression which connects implicit and explicit understandings. I related progression to recent geographical efforts to rethink abstraction in a productive rather than reductive sense, and rendered my own research narrative from Chaps. 3 and 4 in diagrammatic form to work through this idea of progression. Injecting Gendlin's philosophy into this work on abstraction encourages greater sensitivity to the lively potential between verbal and visual renderings of academic content for stimulating new perspectives and sensibilities. This set the stage for a discussion of Gendlin's idea of the crossing of concepts as a means of opening up this connectivity to enable explication from the implicit. I explored crossing in relation to verbal concepts of scale around which debate continues within geography, and in relation to visual concepts through my own paintings. The conceptual is always and already more than conceptual, and concepts themselves can provide access to the implicit in their crossing. The crossing of concepts allows their implicit excesses to be accessed jointly for potential explication, suggesting the development of new ways in which we might engage with and implicitly refine our academic concepts. However, this chapter also raised questions as to how these ideas might be operationalized within geographical research, which formed the focus of Part 3.

Part 3 focused on Gendlin's explicatory techniques, and my own experimental application of methods derived from them, to explore their potential, practicality and pitfalls. Chapter 6 provided an account of these experimental efforts to combine Gendlin-informed interview techniques with video-elicitation techniques, and discussed the seemingly contradictory findings generated through my research in the context of Gendlin's philosophy. I proposed that Gendlin's work provides an alternative understanding of flow, which might relate in a very specific manner to focusing within artistic practice, and that it could accommodate the seeming contradictions in my empirical material, emphasizing Gendlin's insistence on implicit–explicit connectivity. I also suggested that although social scientific employment of video and film as an affective medium frequently draws for support on psychological work on mirror neurons, my engagement with Gendlin's work complicates any straightforward assumptions concerning the role of mirror neurons in intersubjective understandings of affect. In particular, whereas conventional video-based research emphasizes the multi-sensorial aspects of video, a Gendlinian understanding is more concerned with the capacity of video to sustain explicit understanding while simultaneously (re)establishing implicit understanding or felt sense. This chapter underlined the centrality within Gendlin's work not of the implicit in isolation but of the relation and essential connectivity between the implicit and the explicit. It also encouraged the use of both combined practice-based and video-elicitation methods, and the development of more flexible research methods that are tailored to individual participants in research into affect.

Finally, Chap. 7 served as a critical check on my engagement with Gendlin, identifying, elaborating and, in some cases, unravelling potentially damaging weaknesses in my research design. I focused on the modifications that I made to Gendlin's explicatory techniques, and the combination of unsettling participant practices with Gendlin-informed explicatory techniques, as the former threatened to undermine the latter. This chapter reinforced the need for tailored and flexible research methods in efforts to access the implicit, and identified further methodological contributions and implications arising from my research. I suggested that the selection of video clips to be used as stimulus material might best be conducted by participants, and that the unsettling of participant practices might enhance the effectiveness of video-elicitation techniques, functioning as a practical form of crossing. The chapter argued that, despite certain shortcomings in the research design, the more substantive of these are less problematic than they might at first seem and invite further disciplinary interrogation in field research in the spirit of perplexing conundrums with the potential to inform research practice far beyond geography.

The potential contribution that the work of Eugene Gendlin can make to non-representational geography and beyond can be characterized in several ways. His philosophical work provides a whole new terminology, with and through which we can think about numerous concerns and practices, and can contribute to contemporary debates in non-representational geography, geographies of artistic practice and more-than-human geography. His psychotherapeutic work, and specifically his explicatory techniques, can inform both specific efforts within non-representational geography to access and apprehend affect, and visual- and video-based research methods across the social sciences. In particular, these methods, as applied and examined in my own research, encourage reconsideration of our understanding of video, greater specificity with regard to the selection of stimulus material, and the development of flexible and participant-tailored research methods.

Fruitful conceptual areas for further investigation include the broader disciplinary utility of Gendlin's terminology and philosophy, and the development of a Gendlin-inspired geographical lexicon, which might help to address concerns over concepts with spatial, scalar, visual, masculinist or other problematic associations. This might link with the exploration and development of methods for working differently with and through our disciplinary concepts, both verbal and visual. Taking specific concepts as stimuli for an exercise in thinking-at-the-edge might identify new avenues through which to access the implicit, stimulating efforts to articulate from our implicit understanding and invigorating efforts to enhance our capacity to think about images in conceptual terms (Hawkins 2015), although questions around intersubjective intelligibility remain to be addressed.

Methodologically, there is a need to test more extensively and more thoroughly Gendlin-informed research methods, both as originally specified by Gendlin and as modified for disciplinary research use. Specifically, the need for, and value of, training in his original explicatory efforts needs to be established, and the need for proficiency, or equivalence of proficiency, on the part of researchers and participants in both practice-based and video-elicitation methods needs to be resolved. The influence of formal artistic training on practitioner capacity for accessing and apprehending their implicit understanding through their artistic practice, as well as for subsequent linguistic explication, needs investigation. A key issue, it seems, is the extent to which formal artistic training can naturalize practice habits in a manner that either avoids detrimental effects on a person's capacity to access and apprehend affect through their artistic practice, or that enhances that capacity. Related to this, intersubjective and intermedium intelligibility of explicated concepts, both linguistic and artistic, demands examination. While formal training might aid communication of the implicit among those with a shared linguistic or artistic literacy, communication of the implicit might be lost on those who do not share such literacy. Alternatively, terms such as Jane's scruffle and imagery such as Susan's tadpole and mobile phone might be more easily transmitted intersubjectively than we would anticipate, even with those outside our immediate practice milieu.

However, two final proficiency related questions remain. How and to what extent can we determine our relative capacity to access, apprehend and communicate the implicit? How can we establish conditions in which this capacity can be optimized for the benefit of geographical research? Exploring and establishing means by which the articulation between implicit and explicit understanding can be optimized, both artistically and linguistically, seems paramount to any effort to access and apprehend the implicit, and further geographical engagement with Gendlin's work has much to offer in this regard.

This first geographical foray into Eugene Gendlin's work has introduced and illustrated his core terms and ideas in the context of geographical research into artistic practice, making Gendlin's work accessible through empirical geographical examples and providing new ways of thinking about artistic practice. It has reported on and critiqued deliberate efforts to apply Gendlin-informed methods in field research. This has given rise to previously unforeseen implications and recommendations in relation to specifically Gendlin-informed research and to more general research involving practice-based and video-elicitation methods. It makes specific contributions to the geographies of artistic practice by identifying potential opportunities and challenges for disciplinary efforts to work with images and image-making conceptually and critically (Thornes 2004; Hawkins 2015), and to geography more broadly by indicating how visual methodologies might lead us to new types of knowledge and methods to support research into non-cognitive embodied experiences (Crang 2002, 2003; Pink 2012). It has challenged the capacity of Gendlin's philosophy to function effectively in an applied sense to account for varied and surprising research findings, which is both encouraging and illuminating with regard to its potential to stimulate further investigation. Finally, it has characterized particular aspects of my research design and methods in Gendlinian terms (such as the use of comparison between activities as a practical form of crossing), providing further means by which geography can be granted a Gendlinian flavour.

I conclude on a distinctly and appropriately Gendlinian note, borrowing the sentiment of his own words, to invite others to consider this geographical engagement with his philosophy and techniques not as a completed project but as a first attempt that at the very least provides food for thought and investigation, and to "use it in any form whatsoever, or argue with, do anything with it" (Gendlin 2006: 8). It was in the spirit of exploratory engagement that I took the first tentative steps towards a Gendlin-informed geography. It would be lovely to have company for the onward journey.

REFERENCES

Crang, Mike 2002 Qualitative methods: the new orthodoxy? *Progress in Human Geography* 26 647–655

Crang, Mike 2003 Qualitative methods: touchy, feely, look-see? *Progress in Human Geography* 27 494–504

Gendlin, Eugene T 2006 Transcript of Gendlin Templeton Lecture. Psychology of Trust and Feeling Conference. Stony Brook University. http://www.focusing.org/gendlin_templeton.html. Accessed 05 Nov 2012

Hawkins, Harriet 2015 Creative geographic methods: knowing, representing, intervening. On composing place and page. *Cultural Geographies* 22 247–268

Pink, Sarah 2012 Advances in visual methodology: an introduction. *in:* Pink, S (ed.) *Advances in visual methodology*. SAGE, London 3–17

Thornes, John E 2004 The visual turn and geography (Response to Rose 2003 intervention). *Antipode* 36 787–794

INDEX

© The Author(s) 2016
J. Banfield, *Geography Meets Gendlin*,
DOI 10.1057/978-1-137-60440-8